Internet of Things
Approach and Applicability in Manufacturing

Internet of Things
Approach and Applicability in Manufacturing

Ravi Ramakrishnan
Loveleen Gaur

CRC Press
Taylor & Francis Group
Boca Raton London New York

CRC Press is an imprint of the
Taylor & Francis Group, an **informa** business
A CHAPMAN & HALL BOOK

CRC Press
Taylor & Francis Group
6000 Broken Sound Parkway NW, Suite 300
Boca Raton, FL 33487-2742

© 2019 by Taylor & Francis Group, LLC
CRC Press is an imprint of Taylor & Francis Group, an Informa business

No claim to original U.S. Government works

Printed on acid-free paper

International Standard Book Number-13: 978-1-138-59815-7 (Hardback)

Library of Congress Cataloging-in-Publication Data

Names: Ramakrishnan, Ravi, author. | Gaur, Loveleen, author.
Title: Internet of things : approach and applicability in manufacturing /
Ravi Ramakrishnan, Loveleen Gaur.
Description: Boca Raton : Taylor & Francis, a CRC title, part of the Taylor & Francis
imprint, a member of the Taylor & Francis Group, the academic division of T&F
Informa, plc, 2019. | Includes bibliographical references and index.
Identifiers: LCCN 2019010550| ISBN 9781138598157 (hardback : acid-free paper) |
ISBN 9780429486593 (ebook)
Subjects: LCSH: Internet of things.
Classification: LCC TK5105.8857 .R36 2019 | DDC 004.67/8--dc23
LC record available at https://lccn.loc.gov/2019010550

Visit the Taylor & Francis Web site at
http://www.taylorandfrancis.com

and the CRC Press Web site at
http://www.crcpress.com

Contents

Preface

The Internet of Things (IoT) has arrived as one of the biggest changes to disrupt the digital world and in a much bigger way than what the internet did a few decades ago or evolution of computers did half a century before. It seems very likely that the traffic generated by the IoT devices will far surpass the total traffic generated by human beings so far and that there is an imperative need to rethink the way we have designed our information systems and our computing infrastructure. A number of adoptions of the IoT concepts are visible in all walks of life globally and the number is all set to increase to billions of connected objects before the turn of the decade. This book presents some of the use cases of the IoT in different business facets and processes, focusing more on the manufacturing sector and, hence, there is a distinct coverage of Industrial IoT (IIoT) as well. This book is unique as it highlights a variety of topics and practical implementations, all of them of crucial importance for the futuristic objects-driven internet world.

Many of the definitions of the IoT are still not comprehensive but people have tried defining them. Somewhere, we need to revisit whether the IoT will be complete without people, animals, and all things living becoming connected to the objects there by extending to an Internet of Everything (IoE). This book provides a comprehensive overview of the IoT concept starting from the very basic definition and current technologies, moving over to business models that can and will come in play with IoT adoption. It also goes into detail into different business operations such as manufacturing, energy, logistics, and distribution becoming affected by the IoT along with use cases derived from primary data collection or literature review.

The next thing is to gauge the perception of the IoT (domestic as well as industrial) including a model to gauge the maturity of an IoT-enabled organization. It would be very interesting for readers to try and understand what would be the future trends in the IoT; this is what is listed out in another chapter based on insights and focus interviews with experts and inputs from development labs. Security and privacy are one the most significant concerns for technology adoption amongst people and businesses. The recent spurt in attacks relating to information technology had further aggravated this concern and hence an entire chapter is focused on this subject. It seems clear that human beings will slowly be forced to become secondary citizens of the World Wide Web and objects will claim the first position by virtue of their count and data-transmitting capabilities.

We trust that reading this book and its chapters will provide you with a broader insight of the Internet of Things, its adoption, and future trends.

Ravi Ramakrishnan
Loveleen Gaur
India, December 2018

Abbreviations

AI	Artificial Intelligence
AMT	Advanced Manufacturing Technology
BLE	Bluetooth Low Energy
BOM	Bill of Material
CIM	Computer Integrated Manufacturing
CNC	Computer Numerical Control
CPS	Cyber Physical Systems
CRM	Customer Relationship Management
DSC	Distributed Control Systems
DSS	Decision Support Systems
EER	Energy Efficient Rating
ERP	Enterprise Resource Planning
HMI	Human Machine Interface
IAB	Internet Architecture Board
IBEF	Indian Brand Equity Foundation (www.ibef.org)
ICS	Industrial Control Systems
ICT	Information and Communication Technology
IoE	Internet of Everything consisting of people, process, data
IoT	Internet of Things
IT	Information Technology
JIT	Just-in-Time Inventory Concept
KPI	Key Performance Indicator
LTE-M	Long Term Evolution for Machines
MCDM	Multiple Criteria Decision-Making
MIS	Management Information Systems
NB	Narrowband
NFC	Near Field Communication
OEM	Original Equipment Manufacturer
OSI	Open Systems Interconnection
OT	Operational Technology (shop floor machine operating technology)
PLC	Programmable Logic Controller
RTU	Remote Terminal Units
RFID	Radio Frequency Identification
SCADA	Supervisory Control and Data Acquisition
SOC	System On a Chip
SQL	Structured Query Language
TEC	Telecommunications Engineering Center
TDMA	Time Division Multiple Access
VR/AR	Virtual Reality and Augmented Reality
WAN	Wide Area Network

Authors

Ravi Ramakrishnan, PhD, is an acclaimed Information Technology professional with more than 21 years of global experience in the corporate world. As an award-winning Global CIO with a strong technical and managerial background, he has done numerous global rollouts of Enterprise Information Systems—ERP/CRM/BI and M2M/Mobility and IoT solutions that have been widely acknowledged and awarded in different forums. He has submitted his thesis for Doctorate in Information Technology Management with focus on the IoT strategies and technologies and has published papers on the IoT in Springer, Pertanika, and IEEE Xplore and book chapters on the IoT in IGI Global.

He is a Senior IEEE member and has implemented projects in the United States, Europe, Asia, Africa, and the Middle East across multiple cultures, industry domain verticals, and technologies. He did his doctorate from Amity University with focus on IoT in manufacturing, did his MBA at the Faculty of Management Studies, Delhi University, and has further done a Post Graduate Diploma at AIMA in IT systems. He is DOEACC A level certified and has a Postgraduate Diploma in Operations Management from IGNOU after a Bachelors in Science. He is a Prince 2 certified professional, Microsoft Certified Professional, and an Oracle Certified professional.

His global awards include Computerworld 2016 Global 100 CIO honoree (Florida), IDG–CIO 100 India Award winner consecutively 2011, 2013, 2014, 2015, IDC–CIO 100 award winner 2014, 2015, Dataquest CIO award winner in Mobility and the IoT categories, Chief Information Security award winner CISO 2013, 2014, Oracle Best Implementation award winner 2016, Innovative CIO award winner 2015, 2016, C-Change Awards CIOL 2015, Business World CIO 3.0 award winner 2016, Information Week CIO and Business Impact Leader award 2015, Information Week Edge Award 2014, and CIO Power List IoT Icon and Manufacturing Icon Award winner 2017 from Core Media. He has been a key note speaker at many forums including Oracle Openworld on topics ranging from the IoT, the cloud, and mobility.

He also conducts corporate trainings and trainings for professional management programs and mentors startups working in the field of the IoT, mobility, machine learning, and computer vision.

He can be reached by email at ravi.ramakrishnan@gmail.com

 **Prof. Dr. Loveleen Gaur, PhD,** Research Head and Faculty, Amity International Business School, Amity University, Noida, is an established author, researcher, teacher, educator, consultant, administrator, and program leader. For more than 13 years, she served in India and abroad in different capacities. Before Amity University, she worked with various education institutions and worked as faculty and has done volunteer research work at the University of Maryland, USA. She is actively involved in teaching information systems, research methodology, data analytics, and software tools like SPSS, R, and EXCEL to postgraduate students.

Prof. Gaur has significantly contributed to enhancing scientific understanding by participating in over three hundred scientific conferences, symposia, and seminars, by chairing technical sessions, and delivering plenary and invited talks. She has specialized in the fields of information science, the IoT, data analytics, e-commerce and e-Business, data mining, and business intelligence. Prof. Gaur pursued research in truly interdisciplinary areas and has published six books and more than fifty research papers in referred journals.

Prof. Gaur holds various prestigious positions in India and abroad as a member of the Area Advisory Board of Various Business Institutions in India and Abroad, Board of Studies (BOS), and IQAC (Quality) team, and member of Student Research Committee (SRC) and Department Research Committee (DRC). She is actively involved in curriculum development, outcome assessment, and quality of assignment. She is serves as editor for many national and international journals. She is actively involved in various projects of the government of India and abroad. She has been honored with prestigious national and international awards such as "Senior Women Educator Scholar Award" by the National Foundation for Entrepreneurship Development on Women's Day, the "Sri Ram Award" by Delhi Management Association (DMA), the "Distinguished Research Award" by Allied Academies presented in Jacksonville, Florida, and the "Outstanding Research Contributor" award by Amity University. Apart from teaching, she is actively involved in providing talks to reduce the gap between academia and corporate and provides consultation to new ventures.

She can be reached at gaurloveleen@yahoo.com.

1

Demystifying the Industrial IoT Paradigm

1.1 Primer

Internet of Things is considered by many as the most disruptive revolution, primarily driven by the need of organizations and people to be able to follow objects and make them communicate (Lianos & Douglas 2000) in the field of pervasive and omnipresent computing, the biggest technological change the world has ever seen after the advent of the internet. First coined by the British technologist Kevin Ashton in the 1990s, the demand and potential for IoT-connected devices and objects (Ferguson, 2002) and genesis of associated business models has increased multifold. IoT has two distinct yet overlapping connotations, the first being Consumer IoT for which numerous literature exists and the second being Industrial IoT. In this chapter, we will understand what exactly Internet of Things (both IIoT and domestic) is as defined by authentic literature. The Internet of Things blends the physical and digital world, which offers unlimited opportunities, but also faces a lot of challenges in terms of interoperability, limited compute power, power consumption, and ethical aspects of privacy and security. In short, the IoT can be defined as a well-defined network (Nunberg, 2012) comprising of physical objects or real-world devices, moving vehicles that may be autonomous or human driven, architectural places and other daily use objects incorporating electronic sensors, embedded software, and provision for data network connect—allowing them to collect and interchange data (Kosmatos et al., 2011). This arrangement allows objects and devices to be sensed and also controlled remotely using digital networks. The recent years have seen increasing interest and adoption in the field of IoT, powered by technological advances in embedded systems hardware, software, and connectivity. As more and more tiny, cheap, power-efficient microcontrollers and peripherals are becoming available, there is an increased proliferation of a new category of computers: the IoT low-end devices. Most of the IoT low-end devices have enough resources to run newer operating systems and cross-platform application code.

The Internet of Things has shown rapid evolution with widespread technical, social, and economic impact. This has resulted in a paradigm shift, machines taking over the role of human beings when it comes to data generation and usage. It is projected that by 2020, 100 billion connected IoT devices could be in existence (Biddlecombe, 2009).

The concept itself is quite old and has existed from the late 1970s or earlier when telecom networks started connecting to transmit voice data. Remote monitoring of the electrical grid's supply of domestic power has been in commercial use since the 1950s. The 1990s ushered in an era of Machine 2 Machine (M2M) communication (Reinhardt, 2004) between industrial machines, which existed as closed and proprietary solutions. Later on, the advent of RFID-based solutions, both passive and active, were used for extending object's capability to transmit information, although they were not smart and could just transmit an identifier information (Kosmatos et al., 2011). RFID was widely used in logistics movement and for tracking material and inventory (Sun, 2012). Today, the world has moved on to embrace IP-based networks, which still pose challenges for migrating such solutions due to basic design differences.

It would be very relevant to define the Internet of Things and look at some standard definitions.

The Internet Engineering Task Force (IETF) has defined the IoT as:

> The basic idea is that IoT will connect objects around us (electronic, electrical, and non-electrical) To provide seamless communication and contextual services provided by them. Development of RFID tags, sensors, actuators, mobile phones make it possible to materialize IoT which interact and co-operate each other to make the service better and accessible anytime, from anywhere.
>
> **Z. Shelby**
> *News from the 75th IETF, August 3, 2009*

The National Institute of Standards and Technology (NIST) defines the IoT as:

> Cyber physical systems (CPS) – sometimes referred to as the Internet of Things (IoT) – involves connecting smart devices and systems in diverse sectors like transportation, energy, manufacturing and healthcare in fundamentally new ways. Smart cities/communities are increasingly adopting CPS/IoT technologies to enhance the efficiency and sustainability of their operation and improve the quality of life.
>
> **NIST**
> *Global city teams, 2014*

W3C addresses the IoT as a Web of Things:

> The Web of Things is essentially about the role of web technologies to facilitate the development of applications and services for the Internet of Things, i.e., physical objects and their virtual representation. This includes sensors and actuators, as well as physical objects tagged with a bar code or nfc. Some relevant web technologies include HTTP for accessing restful services, and for naming objects as a basis for linked

data and rich descriptions, and javascript apis for virtual objects acting as proxies for real world.

W3C
Web of Services, 2018

IEEE has defined the IoT comprehensively as a three-layer structure comprising of Applications, Networking, and Data Communications and Sensing.

In 2013, the Global Standards Initiative on Internet of Things (IoT-GSI) defined the IoT as:

> The infrastructure of the information society. IoT in combination with sensors and actuators becomes an instance of cyber-physical systems, which also encompasses technologies such as smart grids, smart homes, intelligent transportation and smart cities.

ITU
Internet of Things Global Standards Initiative, 2015

As per the Internet Architecture Board (IAB), the phrase the "Internet of Things" denotes:

> A trend where a large number of embedded devices employ communication services offered by the Internet protocols. Many of these devices, often called "smart objects," are not directly operated by humans, but exist as components in buildings or vehicles, or are spread out in the environment.

IAB
The Internet of Things, 2016

Enterprises can benefit using the IoT from aspects of asset tracking, manufacturing automation, product innovation based on data, and the ability to physically control machine assets, which can further help in dynamic scheduling and real-time monitoring (Moeinfar et al., 2012).

With more stress on limited resources such as motor able roads, potable water, reducing forest and green cover, and an ageing society enabled by medical facilities, the IoT can prove to be the panacea required to economize the consumption, on one hand, while monitoring and prioritizing generation on the other (Butler, 2002).

At the same time, the adoption of the IoT has been fraught with a myriad of issues such as challenges in power consumption, security and privacy concerns, miniaturization yet reliable, roadmap for adoption, especially the cost of upgrades required for making assets and products compatible, and finally the human or emotional aspect of making "things" accountable and giving them control in our day-to-day lives (Arampatzis et al., 2005).

Some issues are more technical such as interoperability, standardization, compute power, and rapid innovation making investments redundant. Acceptability issues driven by the need for a higher "maturity quotient" is another challenge making businesses shy away from an enterprise level adoption and making them focus on a piece meal approach toward implementing the IoT technology enablers.

Today, the landscape is very different from what we foresee to be in store for the IoT tomorrow. The major ICT players like Google, Apple, Cisco, Salesforce, or Oracle, have changed their business models to cater to an expected demand for IoT, while telecom players have started interconnecting their telephony equipment as a core business strategy. The majority of governments, be it Asia, Europe, Middle East, or the Americas have set up a task force and implementation committees to usher in the IoT technology in different facets, be it Smart Cities, Industry 4, or Smart Transportation.

Further scope exists by integration of the IoT with technology trends like Big Data, Fog Computing, Augmented and Virtual Reality applications, wearables, and Cloud infrastructure adoption.

The situation tomorrow may expand the horizons of the IoT to include IoE, namely people, data, and process into a universally connected digital Earth where everyone and everything is just a few network hops away. The challenges will arise from fragmented and heterogeneous technology Siloed applications, vertically closed systems, enterprises closed mindset relating to exposing data, or facilities to collaboration. New ecosystems will need to evolve supported by new business modes, which are more digital focused than physical product focused, driven by large-scale pilot trials and consumer trials and a startup environment that develops peripheral yet value yielding use-case driven solutions for the IoT environment (Chen & Jin, 2012).

The success factors will involve a minimal commonality approach toward the IoT integration architecture development to reduce divergences among target systems, demonstrating value creation not observed by traditional people-centric manufacturing, ecosystem development for the IoT solutions involving policy governance and a positive culture, and a socioeconomic legal framework to define rules for machine interaction and accountability in business environments.

The world is heading toward a convergence of digital and physical and what traditionally was deemed fit will fall out of place in the new digital world order, where pervasiveness and ubiquitous computing will be the watch words. The ability to digitally track measurements, understand from them the failure patterns, and predict the failures in advance using patterns and data has made the adoption of the IoT even more compelling in manufacturing industries.

1.2 The IoT Adoption—Technology or Strategy Decision

Information Technology and associated developments have changed the way organizations do business. From bulk manufacturing to an order size of "one," manufacturing has become more and more customer preferences oriented, even if it comes to a price premium. Automation using the IoT has so far been incremental, evolutionary, and multiple firms have adopted it as either part of a technology upgrade roadmap pushed by sellers and solution providers along with internal pull from stakeholders in business. Be it employees, partners, government regulations, or strategy roadmaps powered by the way competitors have evolved, customer service demands have necessitated a change, or as part of product upgrade roadmaps, providing value to the end consumer. The IoT as an extension of traditional IT, has both strategic and technological dimensions. Strategically, enterprises will be able to reach out faster to customers, provide proactive services and maintenance on their products, and get usage data, which can be used for new product development or for maintaining assets and reducing downtime so as to improve productivity and efficiency in operations, thereby increasing revenue and reducing costs. From a technology standpoint, the IoT provides data on a real-time basis using sensors enabled with capturing data at end points on a continual basis on usage or expiration of shelf life, and covert data such as location, speed, travel time, etc. This data can further help with MIS decisions. Again, such massive data also referred to as big data, requires corresponding backend systems of traditional IT to store and analyze the same. Similarly, data transfer has to happen across communication channels capable of transferring such amounts of data using next generation networks (Lee & Lee, 2015).

With IoT, the trends will shift from IT teams to business users and people as objects trained by means of repetitive data and patterns intelligent work on routine and often mundane tasks with a higher efficiency as compared to humans. Examples of this could be invoice processing based on business rules or sales order processing, purchase order processing, etc. Robotic algorithms based on AI engines like Tensorflo or Neural Networks can help insurance analysts estimate the cost of motor repair based on images and data from collision sensors of the accident to such a high degree of accuracy that it may even be possible to estimate the cost of repairs and get upfront approvals from the IoT systems (Manyika et al., 2011).

All this will succeed only when open standards for Industrial IoT adoption are formulated and verified through numerous case studies and live implementations by experts and scientific bodies. Various government bodies across the globe have set up a task force to formulate the IoT and AI policies and encourage adoption, such as the Artificial Intelligence Task Force set up by the Ministry of Commerce and Industry, Government of India, and the IoT division in TEC.

A major boost for the IoT comes from the fact that we now have a technology using the much awaited IPV6 addressing scheme (Atzori et al., 2010) to assign a unique address to each grain of sand on Earth. Until now, it was difficult since IPV4 had a size limit of address pools, which has been almost breached with limited computing devices connected to the internet such as laptops, desktops, server infrastructure, and handheld terminals. There was no way an additional IP address could be freed up for smart objects and consumable products. Without a unique address in the internet, it was impossible to communicate and the IoT had its basic premise in the ability to send and receive digital information on self and surroundings, so the capability to broadcast and the willingness to broadcast enabled by a digital connectivity using a unique identifiable address was paramount. There is a significant shift in the learning and knowledge life cycle as well—retention, transmission, evolution, and storage. Earlier, the technology progress of centuries and decades was mapped to years as every couple of years, technology becomes obsolete and is replaced with new technology, be it in compute power, sensor upgrades, and storage or communication channels. And with the advent of artificial intelligence (AI), human beings are witnessing a phenomenon where our brains and thought process are now shifting from primitive activities, such as remembering numbers, to advanced activities such as predicting based on data. The next decade or so is forecast to be knowledge driven with huge databases supporting decisions, specialized engines providing us with much needed analytics by crunching data numbers, and self-managed autonomous systems or expert systems. Smart Objects will surround our workplace, the shop floor, transportation logistics, and vital installations.

Whether this will result in a shift toward people or away from people toward objects stands to be seen with time. In the case of manufacturing firms, there seems to be a mixed shift following the same pattern as in the case of industrialization and adoption of IT-based computing. The lower or nonskilled jobs were removed due to automation and industrialization; similarly, data-related nonskilled jobs like recording data from PLC of automation systems or collating this data, or feeding them to a core system like ERP, will get redundant. With two-way communication systems and latest advancements in manufacturing execution systems (MES), it is possible to control the execution of production assets remotely, thereby giving rise to a possibility of real-time planning and scheduling along with queue prioritization based on customer sensitivities and management mandates.

However, MES systems face their own types of challenges in a process manufacturing setup. To understand this, we will deal in summary with the concepts of process manufacturing. The important parameter for the IoT in manufacturing is contextual data and location awareness. Each point inside the works layout of a factory is clearly demarcated during the setup of the factory.

While process or discreet organizations differ in the nature of being continuous operations, where materials change state from one form to another aided by external or assembly lines where individual components are assembled, the main use cases of the IoT can be found in each stage—the receipt of raw materials in stores, the movement of goods across assembly lines or processing machines, the finished warehouse, and finally for offloading items to logistics. In between operations like quality inspection, stock taking, or cycle counting, process planning and issue or materials to works or returns from them also can have the IoT enablement. Support functions such as employee movements, hazardous waste generation, disposal and byproduct handling, discharges and effluents management, energy production conservation, and management and transportation can also be managed using a mix of AI and image recognition riding on a strong AI analytics platform.

The fundamental difference on why the IoT adoption may not just be a technology decision as compared to traditional software like ERP, Product Lifecycle Management, and MES lies in the fact that while these softwares have evolved and originated from their individual areas of application, the IoT will be an add-on layer to connect data from machines to a digital network and hence more of an enabler. The difference from a traditional system where this data was either punched in manually or just expired after a short shelf life inside internal and limited PLC memory, is that such data can now form the base for further prediction and discovery services using AI or can be initially stored for cause-effect analysis in manufacturing systems. Challenges also stem from the fact that most of the machines in manufacturing were never designed keeping this premise of mining data from them in design consideration and hence there are no standards for this data interchange to take place. Either the legacy manufacturing equipment misses this connectivity mechanism with digital networks such as a support for RJ45, Bluetooth, or Wi-Fi, or the data so generated needs a lot of refinement and cleaning with data not following defined formats and also machines not having algorithms built in to generate validated and de-duplicated data. In the absence of data cleansing and sanitizing measures, as well as given the limited memory and hence data storage capacity in machines, the onus lies on digital systems to store the raw data first and then apply transformational and validation rules on the data to generate meaningful updates. The challenges are further augmented as it becomes a business decision keeping in mind not only current practices but a progressively forward-thinking approach on what kind of data fields from this huge repository of machine data needs to be kept and for what period and in what format. It may be advisable to do a tradeoff between keeping raw data at all times at a cost of storage space and compute power, which can be reexplored any time in future or derive analytics and keep it in a summarized state at the cost of losing benefits of deriving new insights in future. Hence, the premise of adoption of IoT, be it a strategic decision or a technology roadmap decision, will largely depend on the organization. Worldwide observations have shown that organizations ranked higher in maturity process or data and driven by the IoT capable leadership takes it up as a strategic decision while organizations which have a strong IT support but may not have an otherwise strong strategic leadership viz-a-viz the IoT will adopt the technology upgrade roadmap route.

1.3 Industrial IoT

The Internet of Things refers to a connected digital network of everyday physical objects connected to the internet and with each other, capable of being uniquely identified and having the ability and willingness to communicate and broadcast data and information about itself and its surroundings, including usage or operating parameters giving rise to something known as ambient intelligence, which is the net sum intelligence of all objects combined together. There exists a wide range of definitions of the IoT and Industrial IoT, which encompass the aspects of a digital nature of objects and machines other than its physical nature along with the ability to connect and transfer data.

The starting premise of an Industrial IoT (IIoT) has been the operational technology (OT), which has existed for many decades and refers to automation and control systems and has, of late, been integrated with digital technology. Definitions of the IoT while focusing on connected devices in the consumer space has rarely fitted well in the industrial phase; however, few literatures have elaborated on this definition.

As a primer, it is important to understand the concepts of Cyber Physical systems (CPS), Supervisory Control and Data Acquisition (SCADA), and Industrial Controls Systems (ICS).

CPS includes systems having physical and digital components. This includes sensors, actuators for capturing data and readings, basic compute power for calculations, and filtering and networking to pass on data to other systems. In real-life cases, a CPS system integrated to a plastic packaging slitter can ring a siren or hooter with lights to draw the attention of an operator or to stop the spinning motor in case of a break in process if observed to prevent scrap generation.

ICS systems refer to different types of control systems and instrumentation needed to operate industrial processes. They are usually proprietary systems running on specialized hardware and software environments that are generally embedded in nature. An example of an ICS system can be the speed controlling console terminal for a running process plant, which can control the running chain speed of a process plant. SCADA, which is part of the ICS, are usually part of process automation systems such as process plants, conversion plants, and gas and oil pipelines and can help in opening and closing valves from a distance or generating or monitoring alarms or meter readings. SCADA systems have two distinct entities: one, the plant or machinery that is getting controlled and second, a set of intelligent devices integrated with the plant or machinery and having some set of sensors or controls thereby measuring and controlling elements in the first part.

IIoT differentiates from a social or consumer internet in the way value is derived. The value in case of consumer internet is generated by way of information and advertisements, while an IIoT generates value by virtue of connected systems, each of which provides input to the other connected subsystems. This way, manufacturing enterprises can use data from different connected machines through their sensors, software, and networks to generate value for decision-making, for example, understanding the machine performance at different running speeds.

1.4 IIoT Taxonomies

A robust IIoT-oriented manufacturing organization primarily contains four subsections: Smart Products, the IoT aware process, skilled people, and a supporting Infrastructure (IT-IS, Assets, and OT). Products which are self-aware can inform other products about their presence or movement in a production or assembly line and directly correspond with their handling machines. People can be connected on a real-time basis sharing online information with devices they are supposed to handle, thereby providing intelligent designs and better quality of service. Bidirectional information sharing in an extended value chain, from raw material supplier to end customer, can help adapt and improvise the process leading to flexible supply chains. Smart Infrastructure interfacing with mobile devices, cloud-based analytics, and products and people can help improve manufacturing efficiency and reduce cost of production.

To understand the IIoT protocols existing today applicable to industrial manufacturing, a basic primer of the 7-Layer OSI model is presented here.

1. Physical layer defines the electrical attributes including the connecting medium such as co-ax or UTP cables of specifications like CAT 6a, 6, 5, and RJ 45 connectors.
2. Data layer ensures reliable delivery with data frame, error detection, and correction, sequence and flow control, Ethernet and MAC are defined at this level.
3. Network layer controls routing and prioritization using address and ensuring delivery to hosts such as IP, PPP, and X.25.
4. Transport layer sequences application data with start and stop bits, including error detection and correction using TCP/UDP protocols.
5. Session layer establishes sessions between applications and networks for control and synchronization.
6. Presentation layer deals with representation of data and coding and ensures data conversion so that it can be transferred across networks.
7. Application layer is used for software to analyze the data for usage by the other layers. Protocols such as FTP, HTTP, and SMTP are found here.

There is a lot of literature existing on industrial focused the IoT from manufacturers and case studies, and generally the following categories of IIoT implementations are observed:

- Device Centric taxonomy (Dorsemaine, 2016) at the basic level captures and provides data about a device, such as measurements and attributes, without any indications of its actual usage.

- IoT stack centric taxonomy (Püschel et al., 2016) covers the complete stack such as the service provided, data, and interaction or functions.
- IoT sensor taxonomy (Rozsa, 2016) is used on sensors which are mobile and can transmit data on position or motion including speed.
- IoT smart environment taxonomy (Ahmed et al., 2016) based on classification of networking elements.
- IoT architecture taxonomy (Yaqoob, 2017) comprising of business architecture and technology.
- IIoT taxonomy (Schneider, 2017) classifying based on reliability, real-time nature, distribution, and collection focus.
- Sector based the IoT taxonomies (Beecham Research, 2014), which relates the device to the business use it is put to, such as manufacturing process plants.

Industrial Ethernet and Industrial Internet both facilitate deterministic ("Determinism means the ability or guarantee by a protocol to ensure messages are sent and received in finite time and with complete packets") controls and increase data access. While the former can be treated as "What" the latter is "How," and hence Ethernet, which has so far proven its robustness in data networks, is now becoming part of advanced industrial manufacturing. A major protocol for industrial Ethernet is "Profinet," and has become the de facto choice for industrial networking. It is an open network and can fit with any existing protocols seamlessly be it HTTP or TCPIP and hence any customer application can ride on it to pull or push data. Profinet enables connecting devices, sensors, subsystems, and systems in a lower cost and more secure manner and brings Ethernet to production assets. It is, however, more robust compared to office Ethernet, which cannot meet the harsh environments of industrial manufacturing environments like dust, vibrations, heat, electro-magnetic interference, and noise.

The PROFINET architecture has the following features:

- Scalability to add number of devices
- Field device accessibility
- Remote troubleshooting, maintenance, and servicing
- Reduced costs

PROFINET is built on three communication services:

1. Standard TCPIP caters to functions such as parameter passing, video transmission, audio transmission, data transfer, and interface with the application tier systems.
2. RT (Real Time) deterministic performance (not observed or required in digital networks) to bring down to 1 ms levels such as motion control values or high-performance motor speed measurements.

3. IRT (Isochronous Real Time) service build on top of ASIC hardware with cycle rates of sub millisecond and jitter rates at sub microsecond. This comes in handy for high-level signal prioritization for high-precision motion control applications.

The other protocols used are EtherNet/IP (leading presence in the United States, managed by ODVA Inc.) and Modbus/TCP (Acromag Inc., IEEE 802.3 compatible).

EtherNet/IP allows classifying Ethernet nodes based on behavior and supports:

- UDP-based data transfers
- TCP-based messaging services: upload and download of parameters
- UDP-based monitoring of state
- IP-based broadcast: all or selective nodes
- Dedicated port numbers for messaging (TCP 44818, UDP 2222)

Modbus/TCP is a Modbus RTU protocol running on Ethernet using TCP interface enabling transmit of data packets from the bus to the compatible devices over Ethernet.

1.5 IoT Challenges in Agile Manufacturing

IoT-enabled work environments can make manufacturing more agile. Agility at individual machine level or mechanical assets could be a result of increasing the set of instructions, permutations, and combinations possible for operating parameters, be it mechanical, software guided, or electrical. Hence, we can customize end products for each of the above combination. Agility could also be achieved by different machines coordinating to create a modular production system, which can then achieve a number of combinatorial factors. These factors can address a myriad of needs or produce different product combinations. This setup can be reused or can be provided on a service model while also being configurable to achieve single lot size production. The challenges in such a setup is driven by the cost of having machines communicate digitally, understand a standardized instruction set from a range of proprietary instruction sets, and thereby achieve an integrated digital and operational network.

The interests of such a scenario in the manufacturing sector arises from the ability to open up assets in a cloud model as a service where asset owners can lease out such assets to support a fully configurable product manufacturing order, creating new business models based on priority queues by making all their devices, machines, and even factories open to distributed computing.

Such a plug-and-play arrangement can help set up virtual manufacturing on the cloud and hence reduce costs of production while improving yield and creating customer multiple choices as compared to a fixed product generated as a result of bulk manufacturing or batch processing. This is one of the key challenges since the IoT enablement in other sectors like Healthcare, etc., has been focused on inexpensive and fairly new generation equipment compared to legacy and capital-intensive manufacturing assets constrained by compute power, memory storage, and a mix of physical and digital controls.

Currently, configuration of new designs in manufacturing assets involve making changes in machine schematics and drawings, creating experimental models, and finally transferring them for mass production of machines; a step which involves numerous engineering steps. This design assumes that the changes or configurations possible are highly restricted to a set of predefined commands and PLC gateways, which are the only source of input and output to the machines from other systems and softwares like ERP or MES.

IoT supported Agile Manufacturing would require a lot less rigid and multiple entry and exit points for the machines, for example, a control to tighten the valves reducing pressure or increase the motor speed or decrease the temperature of melt in a flexible packaging environment.

- *Data on demand*: It should be possible for an Agile Manufacturing system(s) to broadcast their capacity, availability, schedules, processing times, processing capabilities, and product mixes to a prospective user in a cloud environment.

- *Control*: It should be possible for a consumer to customize the material and operating parameters like machine speed, temperature, pressure, etc. This may be for machine level components or individual machines in totality, which should be possible in an autonomous type of operation.

- *Raw Material*: It should be possible to pick up raw material from automated dispensing silos of material in any proportion required and to the extent required backed by further backend supplies of material from connected producers digitally or physically. It should also be possible in case of recyclable material to enable recycling of salvage or process waste back as raw material of same or different grade.

- *Energy*: Requirements could vary based on the process adopted and hence should be supplied on demand.

- *Process Recipes and Formulas*: This being the most important component allows consumers to customize the product mix in terms of process formulas and recipes.

A laboratory example of an Agile factory is the Lemgo Model Factory (LMF), which showcases hybrid manufacturing with process and factory automation elements. The functionality implemented consists of processing corn seeds and producing popcorn over a five-step process.

Currently, in an Agile Manufacturing setup (as opposed to traditional setup, the master PLC execution software of each machine seeks instructions from a consolidated MES and communicates with the input-output devices, actuators or motors over an industrial network bus) the connecting bus gets changed when the connected components are added or removed. This happens in two stages: changes to the bus topology and changes to the PLC program subroutines to enable connectivity. For the above to happen, control software has to be programmed, devices need to be added to the programming language, input and output ports are to be mapped to the PLC, and device details extended to the bus architecture with variable names of ports attached. Hence, this Fieldbus connects the PLC or DCS with field devices such as I/O, drivers, and actuators and improves communications substantially, reducing the challenges faced in adapting the traditional office-Ethernet to a shop floor automation system. PROFIBUS ("IEC61158") is the most widely used Fieldbus and can deliver real-time performance as required for manufacturing asset automation. A significant challenge arises of coexistence of different industrial network protocols, such as PROFINET or EtherNet/IP or Modbus/TCP. Although protocols fully compatible with Office Ethernet can coexist in a single cable, the challenge is more of interoperability, which means only shared usage of infrastructure is possible without any meaningful exchange of data for lesser redundancy of functions. Protocol differences imply that diverse industrial ethernets cannot communicate with each other except with the usage of additional hardware. Add to this the challenge of noncompliant ethernet networks, which are closed networks, and the existence of proprietary networks as well.

Notable is the concept of the "IoT@Work" project architecture, which is a layered process starting from the abstraction layer of devices, management functions, service layer, to the business applications on the other end. Other challenges include security functions, network and communication management functions, and device and service management functions.

1.6 Drivers for IIoT Adoption

The biggest driver for IIoT adoption is development of common standards aimed at interoperability to aid seamless data flow. This can be done by adopting open standards and API as compared to current proprietary standards from future investment-proofing the IoT solutions. The key standards are ISO/IEC JTC1, IEC TC 65, IEC 61987, RAMI 4.0, P2413 Architecture framework, IEC 62541, and MQTT-AMQP. Together, these define the architecture models, process standards, and semantics. The IoT open source initiatives in a consumer-market include OpenIoT, InfluxDB, OpenHAB, IoTivity, and OWASP while for the industrial market includes Node-Red, ROS, and OpenWSN based on Service Layer or connectivity layers separately.

The decision to go for open source or proprietary depends on the following:

- Business case to opt for either one if the machine comes from a vendor and is the primary or same vendor machines in the IIoT network then it makes sense to have a proprietary standard provided by the OEM.
- The ability to have the IoT reference architecture, which is community approved.
- Ability to have agile development and agile go-to market scenarios on a global basis.

Cost reduction has been a main driver for IIoT and hence technology that reduces cost of process or production along with complementary and ancillary technologies, advent of robotics, systems support using Big Data and Cloud, low power, and form factor compute devices have also increased adoption.

To leverage IIoT and the huge data generated to its potential, fog computing is playing an important role, where processing is done at the network edge and too much data transmission does not take place leading to time savings.

Cross-discipline training and multidiscipline approaches for upskilling people is also leading to better data analysis.

New revenue streams are getting generated and new business models are getting envisaged based on real-time production schedules, accurate demand forecasts, JIT and similar delivery models, and MES integration.

A wide range of cloud-based offerings are also becoming an enabler for IIoT adoption, leading to drastic reduction from the prototype stage to the commercial launch for IIoT solutions.

1.7 IIoT for Process Management

Business situations in normal manufacturing versus remanufacturing units are different, while the former are less dynamic but larger in size, the IoT can help with process planning in both types of industries. While remanufacturing helps saves raw material by reusing available material, reducing energy consumption, and minimizing equipment costs, the challenges are in a more complex process planning with fixed suppliers and markets. The IoT historic data can help simulate process planning by improving data acquisition, capturing more accurate data, avoiding human omission, and hence improve data validity. Process controls can be achieved by having data interpreted and validated by subject matter experts locally or remotely, and actionable items so derived to be presented to plant personnel at the right time. With sensor technology advancements including self-powered, self-calibrated, error detecting and correcting, and nonintrusive capabilities production process can send information from hundreds of sensors to monitoring systems including asset management or enterprise resource planning systems through the IoT gateways. Examples can be a propulsion jet motor sending information on temperature and speeds to ground staff, or a liquid packaging slitter plant sending video footage of defects in printing and ink coloration.

IIoT can help improve process safety performance by digitization of risk management and operations management. It can cover the front-end safety design data and assumptions relating to Personal Hazards (PHA), HAZOP, LOPA (or fault tree analysis), and SIL into a digital database that makes safety and design information consistent. Process improvements can be obtained by creating "digital twins," a virtual model of physical assets to allow analysis of data and monitor systems for imminent problems and plan simulations.

1.8 IIoT Protocols

The following layer-wise protocols are applicable to IIoT as well as consumer IoT:

- *Infrastructure focused protocols*
  - IPV6: Internet Protocol Version 6
  - TSMP or Time synchronized network protocol
  - UDP (User Datagram Protocol) more attuned for real-time applications
  - uIP: An open source stack for microcontrollers
  - 6LowPAN IPv6 over Low power Wireless Personal Area Networks working on 2.4 GHz frequency range with transfer rate of 250 kbps
  - QUIC (Quick UDP Internet Connections) with support for multiplexing
  - DTLS or Datagram Transport Layer a highly secure TLS based protocol
  - Aeron with support for UDP multicast and Unicast
  - NanoIP for embedded devices
  - CCN or Content Centric Networks for scalable and distributed architecture
  - RPL IPV6 for low power networks

- *Transport layer protocols and medium*
  - Traditional Ethernet
  - Wireless Hart for process measurement, control, and asset management
  - DigiMesh: A P2P solution used in wireless end point connectivity solution
  - NFC built on ISO/IEC 18092:2004 standards and operating at frequency of 13.56 MHz and a data speed of 424 kbps and short range
  - ANT: A proprietary WSN technology that enables semiconductor radios operating in the frequency range 2.4 GHz used in industrial and medical applications
  - iBeacon and EddyStone based BLE formats for proximity beacons
  - ZigBee uses ISO 802.15.4 standard in 2.4 GHz frequency with 250 Kbps speed and can support a maximum of 1024 nodes with a 200-meter range and also supports 128Bit AES encryption
  - Wi-Fi, LPWAN
- *Data Protocols*
  - Message Queuing Telemetry Transport for light weight messaging with small footprint, an alternative MQTT-SN exists for sensor networks as well. There are various implementations like Mosquitto and IBM Message Sight
  - CoAP or Constrained Application Protocol is used in WSN Nodes with mapping possibility with HTTP thereby extending the IoT to the web. Another variant is SMCP, supporting fully asynchronous I/O
  - Web socket is part of HTML5 and connects client server over a full duplex model
  - NodeJS a JavaScript based library
  - HTTP2 allows efficiency in networks using compressed headers
  - Reactive Streams a JVM based protocol with reduced processing power requirements
  - LWM2M (Light Weight) machine to machine connectivity
  - LLAP (lightweight local automation protocol) similar to short messaging text being interchanged between smart objects. This is being projected as future proof and human understandable
  - DDS or data distributed service for real-time systems
  - AQMP or Advanced Message Queuing Protocol implementing better reliability, message orientation, security, queuing, and routing

- XMPP-IoT (Extensible Messaging and Presence Protocol) is a real-time XML format data transmission protocol for video, voice, and messaging
- M2DA/Mihini agent which mediates between a M2M server and embedded gateway applications
- *Service Discovery aiding protocols*
  - mDNS for resolving name services in small networks
  - Physical web which allows a BLE Beacon to detect nearby broadcasted URL
  - UPnP or Universal Plug and Play to allow the IoT devices connect in the network and discover each other's presence for transmit of data
- *Semantic or definition oriented protocols*
  - IoTDB which follows a JSON linked data standards
  - SensorML provides models and XML encoding for describing sensors and measurement process
  - Wolfram Language enables Device operations like Read, Write, Execute, and Read Buffer
  - RAML: RESTful API Modeling Language for producing reusable code for IoT
  - LsDL: Lemonbeat smart Device Language based on XML-based device language for service-oriented devices
  - SENML for taking readings from sensors and transporting them
- *Device Management*
  - OMA: LightweightM2M v1.0 primarily a device management protocol but can be extended for application management
  - Weave is a communication platform enabling the IoT devices setup and configuration
  - Telehash a secure wired protocol
- *Multi-Layer Frameworks which encompass more than one layer*
  - AllJoyn
  - IoTvity an open source framework developed by Linux
  - IEEE P2413
- *Identification Protocols to identify each smart object uniquely in the IoT world*
  - uCode
  - EPC

- *Security*
  - Open Trust protocol (OTrP), which can manage the application life cycle including installation, updating, upgradation and deletion
  - X.509, which is a PKI standard for digital certificates
- *Vertical Focused*
  - IEEE 1451 for connecting sensors to microprocessors or instrumentation
  - IEEE 1888.3 for security protocols for green community
  - IEEE 1905.1 for Digital Home Networks
  - IEEE P1828 for Systems with Virtual Components

1.9 Product Development and IoT

Market-driven product design and development is now the norm for consumer industries, based on capturing product attributes and processes along with usage data. IoT-based decision support systems can help predict customer purchasing behavior and factors influencing and estimating customer lifetime value and finally understand the customer's sentiments and usage of a product to develop and design a better and more robust product.

The strong business case for IoT-aided Product development comes from lowering of data driven R&D and hence development costs are lower, time to market is significantly reduced, new products are no longer only physical but also digital, and this also generates revenue, increases customer lifetime value by being in touch with customers using digital channels, reduces service costs, and reduces warranty costs with proactive maintenance.

IoT data can provide usage data and conditions on self and surroundings, enabling data-based closed loop life cycle management. It is also possible to integrate PLM systems and product data to create virtual 3D prototypes and reduce physical prototyping. Connected products provide continuous usage information for failure prediction, root cause analysis, and design.

Engineering enterprises are evolving ways to integrate hardware, sensors, microprocessors, and storage for enabling ubiquitous computing and universal connectivity in their devices to transmit operational data using analytics about their mechanical and electrical components. This has huge potential for providing a competitive advantage by evolving different business models and operations. An enterprise can achieve first mover advantage by digitizing their product portfolio since design feedback and customer usability feedback can come almost immediately from the products themselves removing guesswork about consumer needs and providing an empirical decision-making tool. Different use cases of the IoT in products in companies like Coca-Cola (using its Freestyle connected vending machine to predict spikes in beverage consumption), Bosch (health buddy to monitor and alert chronically ill patients), iRhythm (heart-related devices) show how it can help in product positioning, appeasing consumers, reducing operating costs, or even creating new, more satisfying products.

Traditionally, customer interaction was only at the point of sale, and anecdotal reporting using surveys or service calls had limited utility. With real-time communication and data reporting, product performance, customer usage, and repeat purchasing behavior have become a real possibility. This indicates a paradigm shift from sales being the end point to actually the starting point of a customer relation. By understanding how customers are using products, companies can build more robust features, add and remove the useful and redundant features in future models, and observe and act of complaints even before the customer reports them, thereby building high-performance products. With connected devices, the whole value chain from the originating supplier of raw material to the end consumer gets a full visibility and feedback; something that is not possible traditionally. Hence, products can be tailor-made with a batch size of one and highly personalized, thereby getting more acceptability.

1.10 Industry 4.0, IIoT, and Related Developments

The first three phases of the industrial revolution were marked by mechanical production (steam and water), electrical-aided bulk manufacturing, and use of electronic processing systems for automated industrial activity, respectively. The fourth industrial revolution or Industry 4.0 as it's commonly referred, indicates the advent of cyber-physical systems capable of communication with each other and to some extent autonomous processing and decentralized machine-driven decision-making. The concepts of Industry 4.0 originated in Germany and the IIoT developed in the United States have a lot of similarity. The basic premise exploits usage of the internet and IoT, integration of digital and physical processes, digital mapping of the manufacturing world, Smart Assets, and Smart Products.

Industrial Internet in the United States (initiated by GE) combines big data analytics with the IoT and is broader in application area than Industry 4.0 since it also covers power generation, distribution, healthcare and manufacturing, transportation, etc. In France, a similar concept of "Industrie du future" was introduced, built on five pillars: First, making companies adapt new technologies; second, new technologies such as 3D printing or additive manufacturing, IoT, and AR; third, employee training; fourth, international cooperation; and fifth, promoting the industry of the future.

A similar initiative "Made in China 2025" was drafted by a Chinese ministry to adapt Industry 4.0 to China's needs to transform its industry and make it innovation driven. Other considerations are green energy, sustainable development, information technology, and robotics, to transition from low cost to high-quality manufacturing. Industry 4.0 is also required since the traditional profit models of industrial manufacturing are no longer exploitable and benefits range from shorter marketing time, improved customer response, decrease in production costs, efficient environmental management with energy conservation, effluents and discharge management, and better work environments.

The "Smart Factory" model of Industry 4.0 is built on the reconfigurable manufacturing system model, which is the next in the series of evolution from first fixed production lines with fixed tasks (one product model) and then flexible production systems with programmable machines (different products but fixed capacities). The production machines used in Industry 4.0 are autonomous and decentralized with self-decision-taking capacities based on data and machine learning and includes robots and robotic arms that can self-optimize.

The end products are also smart with sensors that can connect through low energy networks for context specific information, localization, and operating environmental state parameters. These may feature built-in capabilities and compute power for processing data and control their logistical path or trigger workflows. This data can be used for proactive maintenance.

Production elements in Industry 4.0 also have a digital virtual identity, are also interoperable and connected, and Human to Machine (H2M) and Machine to Machine (M2M) collaboration is commonly found, thereby splitting tasks earlier performed by human beings between robots and machines. Additive manufacturing or 3D printing enables product design models, thereby shortening the field trials post factory development.

An important component of Industry 4.0 is the IT systems support, which includes ERP, CRM, and PLM, MES and process level controls, and machine level PLC/controller architectures.

A reference architecture model RAMI4.0 was developed to address the challenge of incorporating existing standards into new concept (Figure 1.1).

The meta-model in three dimensions helps in:

- Identification of existing standards,
- Identify and close gaps in existing models, and
- Identifying overlaps

The first dimension addresses two elements, namely instance and type (any product or idea which is still in the planning phase and not available yet). The second dimension deals with location and functional hierarchy between product and other subsystems. The third dimension consists of multiple functional layers as below:

- Asset layer including conveyors, PLC, software, etc.
- Integration layer including digital processed information generated by sensors or ICT controls

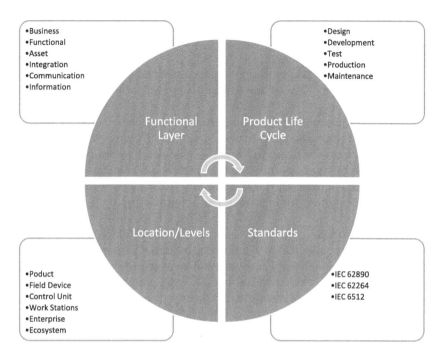

FIGURE 1.1
Smart manufacturing landscape.

- Communication layer consisting of uniform data format and protocols
- Information layer for processing data
- Functional layer, which is a broad abstraction including software functions like modeled in ERP
- Business layer consisting of business models and links between different processes

Finally, the following preconditions are to be fulfilled as part of Industry 4.0 before a new concept is introduced.

- Stability of production to be ensured
- Stepwise investment should be possible
- Security and privacy issues to be addressed

Also, the entire value chain including suppliers, consumers, and all enterprise functions must be included other than the production systems under the ambit of Industry 4.0.

1.11 Conclusion

This chapter has discussed numerous concepts related to Industrial IoT, which is concerned with the use of the IoT in an industrial environment. These concepts lay the broad level base for advanced concepts in the subsequent chapters, which are more focused on individual components corroborated by case studies. For manufacturers to bring together all connected devices, they need to innovate business models and use cases and automate transactions. The biggest challenges are of standardization, power consumption, and miniaturization. Numerous taxonomies in IIoT exists, which have been commercially defined. Smart factories comprising of self-aware and autonomous production systems will optimize, based on resource availability and consumption, across the entire value chain. New product development can be triggered by technological developments, customer demand, market trends, or the ideation process. The IoT has potential for capturing data across all these dimensions. Lastly, the German Industry 4.0 concept and its related variants in different countries was discussed including the reference model and the preconditions, and the three main things we learned are that Industry 4.0 is not only related to production but also to complete value chain from providers to consumers and spans all of the enterprise functions, it assumes support for complete life cycle of systems and products including smart products, which can provide data about themselves and their surroundings, and the IoT basic technology is being applied to the manufacturing sector. The author's view is that a bright potential and unlimited possibilities exist in adopting IIoT in the manufacturing sector, connected objects can communicate with Smart Assets using the IoT compliant infrastructure, and IIoT-driven processes will have better efficiency and less failure rates driven by real-time connected information being transmitted from and between the objects. This can lead to autonomous operations for many functions currently being driven by humans in industrial manufacturing. From process to discreet, assembly to remanufacturing, IIoT can handle all operation aspects including manufacturing, warehousing, material movement, factor maintenance, energy management, and logistics operations.

References

Ahmed, E., Yaqoob, I., Gani, A., Imran, M., Guizani, M. 2016. Internet of things based smart environments: State-of-the-art, taxonomy, and ppen research challenges. *IEEE Wireless Communications.* 23. 10.1109/MWC.2016.7721736.

Arampatzis, T. et al. 2005. A survey of security issues in wireless sensors networks, in intelligent control. *Proceeding of the IEEE International Symposium on Mediterranean Conference on Control and Automation*, pp. 719–724.

Atzori, L., Antonio. I., Giacomo, M. 2010. The internet of things: A survey. *Computer Networks*, 54, 2787–2805.

Beecham Research. 2014. *M2 M Sector Map*. Retrieved from Beecham Research, January 6, 2018, http://www.beechamresearch.com/download.aspx?id=18

Biddlecombe, E. 2009. UN Predicts "Internet of Things." BBC News. Retrieved July 6, 2009 from http://news.bbc.co.uk/2/hi/technology/4440334.stm

Butler, D. 2002. Computing: Everything, everywhere. *Nature*, 440, 402–405.

Chen, X.-Y., Jin, Z.-G. 2012. Research on key technology and applications for the internet of things. *Physics Procedia*, 33, 561–566.

Dorsemaine, B. J. 2016. Internet of things: A definition and taxonomy. *NGMAST 2015 9th Int. Conf. Next Gener. Mob. Appl. Serv. Technol.* (pp. 72–77).

Ferguson, T. 2002. Have your objects call my object. *Harvard Business Review*, 1–7.

IAB. 2016. The Internet of Things. Retrieved from iab, December 15, 2017, https://www.iab.com/insights/connected-devices-internet-things/.

ITU. 2015. Internet of Things Global Standards Initiative. Retrieved from ITU, July 14–20, 2018. https://www.itu.int/en/ITU-T/gsi/iot/Pages/default.aspx.

Kosmatos, E., Tselikas, N., Boucouvalas, A. 2011. Integrating RFIDs and smart objects into a unified internet of things architecture. *Advances in Internet of Things: Scientific Research*, 1, 5–12, doi: 10.4236/ait.2011.11002.

Lee, I., Lee, K. 2015. The Internet of Things (IoT): Applications, investments, and challenges for enterprises. *Business Horizons*, 431–440.

Lianos, M., Douglas, M. 2000. Dangerization and the end of deviance: The institutional environment. *British Journal of Criminology*, 40, 261–278.

Manyika, J., Chui, M., Brown, B., Bughin, J. 2011. *Big Data: The Next Frontier for Innovation, Competition, and Productivity*. McKinsey Global Institute.

Moeinfar, D., Shamsi, H., Nafar, F. 2012. Design and implementation of a low-power active RFID for container tracking @ 2.4 GHz Frequency. *Advances in Internet of Things*, 2, 13–22, Scientific Research.

NIST. 2014. https://pages.nist.gov/GCTC/

Nunberg, G. 2012. The Advent of the Internet: *Courses*. UC Berekley School of Information. Retrieved February 12, 2018 from http://courses.ischool.berkeley.edu/i103/s12/SLIDES/HOFIInternet1Apr12.pdf

Püschel, L., Schlott, H., Röglinger, M. 2016. What's in a smart thing? Development of a multi-layer taxonomy, Thirty Seventh International Conference on Information Systems, Dublin.

Reinhardt, A. 2004. A Machine-to-Machine Internet of Things. Bloomberg Businessweek. Retrieved January 10, 2018 from https://www.bloomberg.com/news/articles/2004-04-25/a-machine-to-machine-internet-of-things

Rozsa, V. 2016. An application domain-based taxonomy for IoT sensors. *ISPE TE*, 249–258.

Schneider, S. 2017. The industrial internet of things (IIoT). In H. Geng (Ed.), *Internet of Things and Data Analytics Handbook*. Hoboken, NJ, USA: John Wiley & Sons, Inc.

Shelby, Z. 2009. News from the 75th IETF, August 3, http://zachshelby.org.

Sun, C. 2012. Application of RFID Technology for Logistics on Internet of Things. *AASRI Procedia*. 1, 106–111. 10.1016/j.aasri.2012.06.019.

W3C. 2018. Web of Services. Retrieved from W3C, April 13, 2018, https://www.w3.org/standards/webofservices/.

Yaqoob, I. 2017. Internet of things architecture: Recent advances, taxonomy, requirements, and open challenges. *IEEE Wireless Commun.* (pp. 10–16).

2

The Internet of Things Applications

2.1 General

The Internet of Things (IoT) is all pervasive and therefore can be used anywhere from personal health tracking wearables to manufacturing and retail (Biddlecombe, 2009). It is also evident that some domains will mature earlier and benefit as compared to others (Lianos, 2000). The IoT application is enhancing our lifestyle, which in turn effects both the domestic environment and the industrial environment (Ferguson, 2002). Industrial manufacturing has its own set of challenges due to environmental regulations, increasing pressure on sustainability, expectation to reduce cost of manufacturing, and ensuring safety and compliance. The IoT has a role to play in all types of industrial manufacturing, be it automobiles, telecom equipment, process plants, refineries, heavy engineering, consumer electrical goods, or pharmaceuticals. With the growth of pervasive computing and unlimited storages, machine data is now becoming evolved, driven by machine learning to provide much needed manufacturing insights. Computer-aided manufacturing has long been in existence and has been instrumental in ushering in an era of lean manufacturing. The high-level business problems addressed by the IoT (Lombreglia, 2010) include increasing productivity through data analytics, increased utilization of assets, supply chain optimization, better bottom lines, and better compliance.

The IoT adoption has been, of late, aided by a number of factors listed below:

- *Pervasive Connectivity*: With data network availability across the globe and with a high degree of uptime, it has become possible to reliably and speedily transfer any volumes of data across digital networks. Low cost and higher resilience have ensured mission-critical decisions can be taken on data on a real-time basis (Butler, 2010).
- *Low Form Factor Computing*: Manufacturing advances have led to the computer-in-a-chip model, thereby helping develop wearables or ingestibles with enough auxiliary power and compute power to tie up with sensors for capturing raw data and transmitting them

to nearby networks. This has given rise to concepts like personal area networks or wireless body area networks creating new cases in medical science applications based on the IoT (Emil, 2005).

- IPV6: This version of the Internet Protocol as against the previous version IPV4 (32-bit addressing scheme, e.g., IPV4 10.10.10.1) was developed by IETF to address the paucity of IP addresses or unique identifiers, which could be given to independent devices to identify them on the World Wide Web. IPV6 uses a 128-bit address and hence allows up to 2^{128} addresses (Aggarwal & Das, 2012). The other benefits include route aggregation, interoperability with IPV4, and higher security. A sample IPV6 address consists of eight groups of four hexadecimal digits separated by a colon, for example, IPV6 2001:0db8: 0000:0042:0000:8a2e:0370:7334

- Cloud Computing refers to a development of providers creating a pool of resources—such as servers, storages, database, applications, software firewalls, network equipment, or even desktops that can be provisioned for end user usage at a short notice. The duration of usage can vary from a few hours to years and a digital connection on the internet (public cloud) or intranet (private cloud) is the only thing required to assess this resource.

- Big Data Analytics refers to the examination and processing of large volumes of data which may be structured or unstructured (big data) to extract hidden patterns and identify trends such as customer preferences or sales volumes or machine performance (Razzak, 2012).

- Open Standards refers to a deviation from earlier proprietary standards so that heterogeneous machines can speak and interact with each other using a common set of protocols or semantics, thereby exchanging data for a combined data visualization across the extended value chain.

The McKinsey Global Institute describes a variety of potential business application areas that will derive value from connected objects. This is partly due to the ability of the IoT to unify and also identify the real-time state of connected objects and partly due to evolving business models, which have a high degree of customizability. As above mentioned, connecting millions of objects is a technical possibility with the development of the IPV6 protocols, thereby assigning unique identifiers to everyday use objects such as clothes, shoes, and furniture, and living objects such as cattle livestock, other animals, and trees (Figure 2.1).

Gartner's information technology Hype Cycle model (Gubbi et al., 2013), which measures in the form of a curve the emergence, adoption, maturity, and impact on applications of specific technologies, has constantly ranked the IoT in the emerging sector since 2012, but it is clear that for the IoT to be

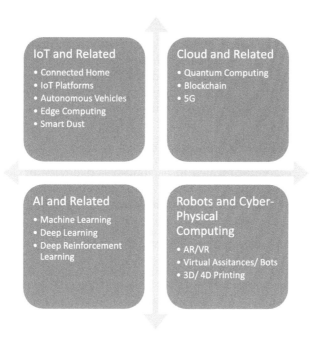

FIGURE 2.1
The next generation high potential technologies (Adapted from Gartner, 2017. https://blogs. gartner.com/smarterwithgartner/files/2017/08/Emerging-Technology-Hype-Cycle-for-2017_ Infographic_R6A.jpg.)

commercially deployable, it must be supported by an ecosystem of sensors, networks, gateways, applications, and communication and computation technology. The IoT has leveraged on some of the proven older technologies like RFID and NFC such as their ability to be recognized wirelessly over short distances (Gigli & Koo, 2011). The oldest of technologies using imaging, such as the barcode, have proven that detection of objects and communication based on a unique identification is possible; however, these were limited to a passive role. Wireless Fidelity (Wi-Fi) adoption has increased potential for wireless sensor networks (WSNs), which connect the IoT devices. ZigBee is one of the protocols for enhancing the features of a WSN due to its low power consumption. Hence, WSNs have been widely used in the fields of homeland security, military, new age agriculture such as precision farming, controlled irrigation, forest and environmental monitoring, manufacturing, and detection of environmental calamities like earthquakes and tsunamis (Dodson, 2008).

2.1.1 IoT Reference Model

To understand the application areas of Industrial Internet of Things (IIoT), we are illustrating the four different types of communication models used by the IoT devices.

2.1.1.1 Device-to-Device Communications

This is the most basic model where two or more devices communicate with each other directly, similar to a Point-to-Point connection. A simple example could be two mobile phones connected by Bluetooth or wearable IoT devices connected in a Personal Area network. Protocols such as ZigBee, Bluetooth, and Z-Wave are used in such a communication.

2.1.1.2 Device-to-Cloud Communications

The IoT device communicates to a cloud-based device or service using an internet connectivity provided by Wi-Fi or wired connection using an application server that can manage data traffic.

2.1.1.3 Device-to-Gateway Communications

This model builds on the IoT gateway, which is a device or a software connecting the cloud and sensors. This is used in personal healthcare tracking devices where the gateway becomes a mobile smartphone running an app and communicating with Bluetooth.

2.1.1.4 Back-End Data-Sharing Model

This model provides a cloud-based revenue model as a service, and can also exist as a combination from other cloud services, the emphasis being on data and its utility.

The IoT reference model derived from the standard Network OSI layer comprises of the following component layers (Gershenfeld et al., 2004).

- Level 1 is the physical layer. It comprises of sensors that may be integrated inside products, machines, or endpoints. These sensors record parameters about self and surroundings. These may or may not be remotely controllable but can definitely transmit data of their measurements such as temperature or pressure.
- Level 2 is the communication layer which ensures reliability and timely transmission of data captured by the physical layer. There exists significant challenges at this layer as not all communication can be IP based, and there will be proprietary systems as well which can be heterogeneous and not compatible. Protocols, routing, and switching are handled at this layer.
- Layer 3, or the data elements layer, converts data received into processed information. The basic requirement of this layer is the huge volume of the data that gets processed at this end. Hence, processing has to be done on a per packet basis, which gets segregated and aggregated and then filtered.

- Layer 4, or data storage, deals with event-based data at rest. This gets stored as a SQL or NoSQL-based big data. This data can be further used by applications for analytics.
- Layer 5 deals with data aggregation from different sources, reconciling data, and storing in accessible format.
- Layer 6 deals with applications like ERP, CRM or industrial applications or mobile applications.
- Layer 7 includes processes and people and involves workflow and data exchange in a multiuser environment.

2.1.2 Current Areas of Industrial IoT adoption

2.1.2.1 Improving Energy Management

Energy is a major component of any manufacturing component since machines are mostly electricity driven. Energy costs significantly affect the bottom line and constitute a fair share of the overall operating expenses of manufacturing. This is higher in heavy engineering such as steel works. Energy sources include state electricity supply or captive homegrown power units based on furnace oil, diesel, natural gas, and, to a small extent, solar or renewable sources.

IIoT can help reduce energy consumption while also keeping track of energy requirements on an ongoing basis. Energy Management using IoT sensors can help detect machine level energy consumption, correlate the changes to machine breakdown trends, and also eliminate wastage by shutting off power when not required based on motion sensors. The IoT-based actuators and motors can also optimize the efficiency level of solar panels by automatically aligning to sun movements. The biggest advantage is of real-time monitoring of energy requirements and consumption rather than a post reporting in the form of power bills. This reduces unbilled energy and also reduces costs proactively.

2.1.2.2 Increase Process Manufacturing Efficiency

Process manufacturing refers to using a process plant to take in inputs and using defined formulas and recipes to generate outputs in the form of main products or by-products, generating effluents and wastages in the process where the net sum of inputs should be ideally equal to the net sum of outputs. It is typically used in petrochemical refineries, chemical industries, food and beverages, and the consumer packaging industry. The ideal formulas and recipes for bulk manufacturing stays the same and, generally, the output of the previous stage forms the input for the subsequent stage.

In a process manufacturing organization, the shop floor consists of one or multiple process plants, each of which requires specific inputs of man, material, and energy, does a batch job processing using the inputs, and generates outputs. The sum total of inputs and outputs always matches taking into consideration invisible laws, by-products, and wastages. An IIoT application in operations helps in monitoring and tracking in order to bring efficiency, prevent wastage, and ensure compliance in any batch processing operation. The IoT-based inventory management solutions including RFID (Want, 2006) or BLE beacons can help manage inventory without human intervention by transmitting data on a real-time basis relating to locations of inventory items, broadcasts on expiry dates, and even assist robotic arms to identify material based on specific SKUs for packing and shipping.

2.1.2.3 Organizing Discreet Manufacturing

Discreet manufacturing industries such as automotive, aerospace, defense, and engineering are commonly plagued by rapid innovation cycles leading to shortened product life cycles, customer service demands, intense competition, a highly fragmented and component based BOM, and finally resource volatility. In Shop floors, connected machine assets communication with each other for workload alignments and connected products sharing information about their movement and locations have now become a reality is large automated warehouses. IIoT can help in improving innovation using sensor-enabled end products that can eliminate or reduce product failures while also giving real-time feedback on product performance parameters and usage from customers. The two major IoT use cases are strategic asset management for preventive and predictive maintenance, and enhancing customer experience to provide more flexibility and customization. These can provide valuable data about user preferences.

2.1.2.4 Safety in Mining Operations

The mining industry is undergoing a lot of innovation and improvement and mine safety and productivity is being enforced using IIoT. Since the mining environment offers some rough working conditions, automotive maintenance and operation of machines helps OEM monitor equipment and ensure workability in this situation. IIoT can help drive agile new processes and associated new revenue models. IIoT can help implement better traceability and improve visibility, ensuring maximum efficiency and controlled variances.

2.1.2.5 Improving Agro Production

Precision farming is getting higher adoption nowadays with the global population projected to touch 9.6 billion by 2050, extreme weather conditions due to global warming, and environmental impact and sustainability issues with traditional farming techniques. Smart farming enabled by the IoT techniques eliminates waste and makes it sustainable commercially as well as environmentally. The IoT-enabled farming helps monitor the soil content, moisture levels, luminosity for multilevel farming, temperature for seed germination, and timed/precise irrigation methods. This data can be further fed to a central monitoring team to provide corrective actions such as supplementing nutrients in the soil or regulating water supply. Remote cameras can also help monitor growth of plants while autonomous vehicles can help with the cultivation process. Crop Metrics is a leader in this area with techniques for variable rate irrigation (VRI), soil moisture probes, thereby customizing water flow depending on soil type and its ability to retain water. Drones with high resolution cameras are now used for crop health assessment or even irrigation and crop spraying. Precision hawk is a firm specializing in Agro-based drones. The IoT applications for livestock farming or cattle farming is also becoming widely used, it helps locate cattle while also checking the health parameters of cattle to prevent the spread of infection. Greenhouse and organic farming have also adopted the IoT for smarter, more controlled operations.

2.1.2.6 Environmental Compliance

Today, manufacturing is subject to a number of compliances in effluents, discharges, waste product recycling, and the treatment of outbound material subjective to local by-laws. Currently, these compliances are enforced manually by ad hoc checking and observations. However, with the IoT devices generating real-time data, it is being more and more advocated to ensure compliances are backed up with data. Industrial production is heading more toward green manufacturing and IIoT can help build sustainability on three fronts: resource efficiency, sustainable energy development, and bringing in transparency. The United Nations (2012) conference defined a target of low carbon, resource efficiency, and socially inclusive economy. With IIoT integrating bridging the OT-IT gap, it is possible to continuously monitor and steer production processes. It is important to ensure that while greenhouse gases may have been reduced in one part of the supply chain, it should not lead to a corresponding increase in other parts of the supply chain. This can be ensured by having holistic data across the supply chain, thereby ensuring transparency. IIoT along with additive manufacturing has a potential of reducing the number of parts and subparts and also making long transportation hauls redundant. As per the Global e-sustainability initiatives estimates, digitization and the IoT can reduce overproduction while also saving water to the tune of 81 billion liters, plus significant electricity.

Energy conservation can also be brought about reducing machine time in standby mode, which may be as high as 30% (Energieeffizienz in der Produktion, 2017). Experiments have shown that it could be possible to reduce energy consumption without making changes in design or reducing production, by just using process efficient and optimized applications.

IIoT can help increase the share of renewable energy consumed in industrial production and also reduce volatility of renewable energy systems. Cyber-physical energy systems and virtual power plants bring in more energy-saving alternatives to a digital manufacturing setup.

2.1.2.7 Retailing and Consumer Analysis

Stores and brands including Lord & Taylor or Hudson's Bay use the technology of iBeacon from Apple integrated in a mobile platform for mobile marketing Swirl. This helps them deliver promotional messages highly personalized for each of their customers with the service delivered using the mobile app. This interaction provides a lot of data insights about the in-store presence and experience. Sensors are dotted along customer movement pathways and this helps store managers better plan store layouts or merchandise placement. In the case of Hugo Boss, heat sensors in clothing stores are used to trace a customer's navigation and high selling or profitable and lucrative products are placed in the path of high movements.

2.2 Emerging Areas of IoT Adoption

2.2.1 Robotics-Driven Warehouses

Benefits of integrating robots for supply chain efficiency and inventory movements have been observed in Amazon, which uses its Kiva range of 30,000 plus robots for servicing its "Prime" customer orders, thereby automatically notifying the status of transit and ensuring zero-defect delivery and fulfilment. Logistics automation has resulted in significant savings of time and costs and space, bringing about a go-go supply chain concept. Another case is Aethons Inc., which has introduced its TUG robots to deliver medication, samples, surgical supplies, and dispose of medical waste with a high degree of reliability and ease of use by automated aerial drones, guided vehicles, and legged robots.

2.2.2 Process Safety

This can be highly ensured by adopting a right mix of IIoT technology. In combination with an image recognition algorithm, it is now possible to see if any manual process compliance is not happening, which may be as simple as not wearing safety gear while working. High hazard industries require monitoring at each step to ensure accident-free working hours. IIoT technologies involving sensors connected to machines can quickly gather data leading to rapid decision-making. It helps in risk minimization by predicting outages or avoiding misses and thereby ensuring production targets are met.

2.2.3 Deep-Learning-Based Predictive Analytics

Deep learning refers to the use of neural networks similar to the human brain concept. These can learn over time with datasets referred to as "training" and then predict outcomes on future datasets. By way of building mathematical models, it can be ensured that critical assets have reasonable uptime, thereby increasing profit margins. This may not be possible by just storing low level data on operating parameters like temperature, pressure, or vibrations. Deep learning can provide the capability to predict the future of the equipment under study. Strategically, big data is important but a more stochastic processing of low-level data is also important to obtain results.

2.2.4 Machine Learning for Decision-Making

Algorithms help solve computational problems that are prevalent in the industrial scenario, and have a lot of emphasis on efficiency, such as Hirschberg's algorithm for DNA and protein synthesis or algorithms for optimized search or sorting or computing a what-if situation with "n" variables. Datasets generated by machine usage, consisting of diverse elements, can be used by machine-learning-based methods, which can give results with high precision and high coverage. Machine learning when applied to the IIoT can have deeper impact than in consumer applications resulting in cost savings, higher productivity, and a larger uptime, especially in asset-intensive and data-rich industries. Machine learning usage can help uncover the underlying trends in big-data sets, such as normal working, exceptional working, and boundary conditions, which are otherwise hidden and cannot be manually analyzed. Examples can be identifying the impact of motor vibration on product stability or correlating temperature changes to a valve functional life. The process involves using machine learning or training it for normal datasets and applying it to a wider dataset to identify exceptional conditions or anomalies. The more the data volume to be analyzed, the higher accuracy of predictive models. A major difference between traditional methods like business intelligence and machine learning is while the former deals with insights from data, the latter focuses on outcomes and training a computer to uncover causal factors and adjust the same with time as more data is analyzed.

2.2.5 Connected Machines Assets

Connected machines further lead to support automation, reducing the involvement of human elements while enabling self-diagnosis of problems and reporting them with supporting data over connected digital networks. The IoT devices fitted into end products can raise auto alerts for replenishment requests (Smart Bins and Smart Racks) to vendors, thereby decreasing service lead times.

2.2.6 Inventory Traceability

Inventory management including warehousing, helps improve productivity by reducing the costs of inventory, streamlining purchases, and reducing the time it takes to locate items. Development of economic and faster technology for efficient warehouse management has resulted in faster inventory management operations; however, this increases the complexity of operations and management control. Traditionally, items used to be equipped with a barcode label, EPC code (Graham & Haarstad, 2011) or more advanced RFID (Kosmatos et al., 2011) (passive or active) tags which used to transmit information either through a manual scanning process or through auto detection using different frequency RFID readers. With the IoT sensors that supplement either RFID or BLE location beacons, it is now possible to send scores of other parameters such as temperature, humidity, pressure, location provider GPS sensors, and others to identify and pinpoint the exact location and storage or the ambient conditions of inventory items along with timelines of movement. This will result in fewer shortfalls, avoiding stock-out or overstocking situation. It also answers, using derived data, questions such as what route and time of the day would be more suitable for transport of perishable goods, what is the quality of food material when it is transported in controlled temperature environments, and if there are deviations in standard storage conditions, all in an auditable trail of data generated by the IoT sensors.

2.2.7 Defective Product Identification

Defective products carry enormous liabilities and safety risks for the consumers; however, with the IoT sensors in a chip it is now possible to detect discoloration, pigmentation, or change in chemical composition, or malfunctioning of products which have a high risk such as safety equipment such as airbags in cars or medical equipment. High speed camera equipment can generate data on visual properties, which can be further subject to artificial intelligence (AI) to identify patterns or abnormalities not visible to the human eye. This can help in ushering a risk-benefit-based defect classification rather than a standards-based classification, for example, changing airbags every two years versus changing airbags when identified as defective using proven sensor technology. The use of the IoT sensors along with AI to postulate whether a product is defective can help with a lot of monetary savings and decrease the risk proposition.

2.3 Worldwide IIoT Adoption Case Studies

This section deals with the application of IIoT in industries across the globe to help understand their role in business transformation. These generally fall under one of the three categories: (1) Smart Manufacturing for monitoring of assets and operations, (2) connected end productions, and (3) connected supply chain.

- JCB India, a leader in the agro machinery and construction segment, manufactures heavy earth-moving equipment and tractors in its five manufacturing locations. The company embarked on a program to connect these machines with the customers to transfer information about the machine operating parameters and location. The objectives were to correctly estimate the health of the machine, its operating time, the fuel spent, etc., to offer the product as a service and ensure revenue recognition. Additional consideration was to ensure security by tracking location and extending it to a geofencing mode.

- Uflex Ltd, a leader in flexible packaging manufacturing, implemented IIoT in numerous areas such as inventory management and plant data analysis. Plants across the globe were connected using PLC/SCADA to push data into digital ERP systems, and this data was further used for insightful decision-making and predictive analysis. The data could be used for breakdown prediction, analyzing faults from machine data, and also in locating inventory item locations on a real-time basis using Bluetooth transmitting beacons.

- New Holland (Fiat) India launched the GPS/GPRS technologies on its tractors under the Sky Watch program enabling remote management and monitoring of its vehicles from a centralized command center. This has led to improvement in service levels and higher productivity leading to product as a service business model being introduced. Tractor owners have now been enabled by this data, started leasing out their tractors, and benefited from getting data on performance parameters, running costs, and maintenance during this let-out period.

- Videocon has enabled its end products with the IoT capabilities. Their Wi-Fi range of air conditioners can be enabled by remote control using a smartphone from anywhere fitting into a larger picture of connected homes. An inbuilt sensor, which is an energy meter, helps conserve energy by tracking power usage and switching the appliance on and off based on user proximity.

- Connected products, especially in the consumer automobile segment like Mahindra Reva car, which is the electric model, and Tesla range of electrical cars, have significantly saved service costs by providing monitoring data of the products.

- Another implementation of the IoT in Bharat Light and Power has made power generation more optimized by enabling two-way communication; thereby, wind turbines can be made to function as a demand of power at a particular time of the day.

- Daimler, a premier manufacturer of cars like Mercedes, has launched the IoT-enabled concept called car2go, which are eco-friendly cars that can be reserved by customers using a smartphone app. Using wireless sensors for monitoring individual vehicle performance, it is now possible to remotely analyze data, and provide an accessible network of vehicles to its customers. An intuitive mobile app allows members to take any of the car2go vehicles distributed around them, or reserve a vehicle for future use. Hence, customers can now use cars, without requiring them to purchase a vehicle or pay for a parking spot.

- Whirlpool implemented the IBM Watson cloud to store data from appliances, thereby providing customers with more tailor-made and customized product and service offerings, adding predictive maintenance capabilities and ensuring customer satisfaction.

- Ericson Maritime ICT has implemented IoT to monitor the cargo between shipping ports. The shipping industry being subject to fragmented supply chain, this was in response to the challenge to ensure timely shipping and overcome logistical challenges. Sensors are used to monitor vessel location, speed, and temperature of refrigerated contents ensuring visibility of cargo from production warehouse to end destination.

- A leading packaging film manufacturer with chemical treatment operations adopted IIoT to reduce quality degradation. The hypothesis was based on a co-relation of machine parameters with the quality of the produced output. Degraded output was a cause of concern due to resulting customer dissatisfaction, sales return, and increased cost of recycling the plastic. The manufacturing process involved winder and slitter as two production plants, and the solution kept track of machine settings parameters so that defects could be tracked historically and co-related with quality failures. The benefits were real-time process visibility.

- Prayas Energy group, India, has implemented the IoT for improving efficiency, optimizing resource usage, improvement in governance, and had a pressing case to check the IoT feasibility. Authenticity of data, transmission in low/sparing/no signal areas for offline storage, and retransmission on signal availability were the key challenges that were taken care of using customized smart meters with IoT capability and transmission using GSM SIM cards.

- A multibillion-dollar global metal forging conglomerate adopted Smart Manufacturing in the direction of Industry 4.0 by integrating its robots, conveyor belts, and machines, each of which has their own SCADA systems due to heterogeneity. These machines were earlier disconnected, and data was kept in isolated silos. With an integrated system, visual analytics and production visibility helped in improving overall efficiency.

2.3.1 Current and Future Areas of Domestic IoT Adoption

Totally different use cases and business models (Shao, 2009) are observed in the domestic adoption of IoT, which are tailormade to capture information about the end consumer behavior, for example, the type of digital content the consumer sees or to capture information about the consumer and surroundings such as their lifestyle pattern and food consumption habits. The traditional IoT is formed from three distinct layers: a perception layer (such as sensors that collect raw data such as pulse rate), a transportation layer (such as IoT gateways using protocols ZigBee, Bluetooth, LTE, BLE), and an application layer (which constitutes the end use such as Smart Traffic management or Smart Transportation).

2.3.2 Cashless Transactions

The transition to a cashless economy has already happened powered by smartphones and digital wallets; however, these still require human intervention. The next wave of consumer IoT will be powered by smart appliances integrating directly with banks and e-wallets for a limit-bound payment initiated by themselves. A common use case can be the replenishment of stocked items in our household refrigerator. The refrigerator can identify objects such as milk bottles powered by high resolution cameras and AI, and then monitor their in-out movements and order supplies using the IoT gateways and even authorize payments on delivery to designated inbound racks. The same can power departmental stores and avoid the hassles of long payment queues. As the world moves away from cash transactions to plastic cards, the IoT has a huge potential as payments become authorized by objects sending secure information to the banks based on consumer acceptance. A hypothetical case would be one where a person picks up a consumable item, for example, a can of soft drink with an embedded low-cost fingerprint reader and drops it in their smart shopping bag fitted with sensors. The fingerprint is the authentication to dial up to the bank to hold funds for payment or reverse the charge back based on whether the object drops in the "smartbag" when the consumer finally exits the shop premises. Although still hypothetical, it is very much possible due to developments in biometrics, making mini-readers much more affordable and smartbags reading tags based on near proximity of 1 to 20 centimeters. Such an arrangement would make long queues redundant and also make billing counters a thing of the past with people no longer manning the use of plastic cards and consumers struggling with bills and credit card passwords.

2.3.3 Retail Industry

The retail sector has immense potential and benefits for both customers and retailers right from inventory management, smart vending machines, and protection against pilferage or counterfeits. Another use case of the IoT in the consumer retail industry is where shopping can be transformed into a contextual aware and fully automated experience. The IoT system using RFID tags or BLE beacon tags can help inventory management and ensure right inventory levels. Smart tags attached to the products allow them to be tracked in real time so that the inventory levels can be determined accurately and products that are low on stock can be replenished. Smart bins can keep a real-time track of the weight of their occupants and their counts can be maintained by machine counters. Smart payment solutions such as contactless payment powered technologies such as near field communication (NFC) and Bluetooth can be used for payments in the retail shops by just tapping mobile phones. NFC is a set of standards for smartphones and other devices to communicate with each other by bringing them into proximity or by touching or tapping them to initiate a transaction. Customers can store the credit card information in their NFC-enabled smartphones and make payments by bringing the smartphone near the point of sale terminals. NFC can be used in combination with Bluetooth, where the role of NFC is to initiate initial pairing of devices to establish a Bluetooth connection while the actual data transfer takes place over Bluetooth. Another development is of smart vending machines connected to the internet, which will allow remote monitoring of inventory levels, elastic pricing of products based on demand and consumption, promotions and advertisement content on the fly, and contactless payments using NFC. Smartphone applications that communicate with smart vending machines allow user preferences to be remembered and learned with time. This data can come in handy to make suggestions to customers based on their buying pattern or just as inputs for new product development or new market feasibility. Smart vending machines can communicate with each other, so if a product is out of stock in a machine, the user can be routed to the next nearest machine. For perishable items, the smart vending machines can reduce the price as the expiry date nears.

2.3.4 Home Automation

Smart lighting achieves energy savings by sensing the human movements and their environments and controlling the lights accordingly (Li & Yu, 2001). The IoT can help residential power management using a combination of smart meters (Sanjana et al., 2016). Smart lighting includes smart meters, luminosity detection systems, solid state lighting (such as LED lights), IP-enabled lights, natural light detection, and adjusters. Wireless-enabled and internet-connected lights can be controlled remotely from IoT applications such as mobile or web application.* It is possible to have a controllable LED lighting system that is embedded with ambient intelligence gathered from a distributed smart WSN to optimize and control the lighting system to be more efficient and user-oriented. Another utility is in the field of security. Home surveillance systems have evolved from static cameras to remote-controlled cameras with tilt and zoom operations to motion-triggered camera recorders. These use the IoT in a big way and it is possible to alert customers

---

* Energy-aware WSN with ambient intelligence for smart LED lighting system control.

of abnormal motion detection inside their homes or office premises. Home intrusion detection systems use security cameras and sensors to detect intrusions and raise alerts on SMS or email or even send images to the nearest law enforcement agencies.

Smart appliances make management easier and provide the status information of the appliances to users remotely; for example, a smart washer/dryer that can be controlled remotely and notify users when the washing/drying cycle is complete.

Smoke detectors are installed in homes and buildings to detect smoke, which is typically an early sign of fire. It uses optical detection, ionization, or air sampling techniques to detect smoke. Gas detector can detect the presence of harmful gases such as carbon monoxide (CO), liquid petroleum gas (LPG), and even air purifiers can now detect abnormal levels of contaminants and trigger off air cleaning.

2.3.5 Energy Management

The biggest use of IoT is now to design and implement smart grid technology, which provides predictive information and recommendations to reduce unbilled energy by keeping a real-time track of the load requirements and thereby channelizing energy to the right load sources. It helps suppliers and their customers on how best to manage power.

Smart grids collect the data regarding electricity generation, electricity consumption, storage, and distribution and equipment health data. By analyzing the data on power generation, transmission and consumption of smart grids can improve efficiency throughout the electric system. Storage collection and analysis of smarts grids data in the cloud can help in dynamic optimization of system operations, maintenance, and planning. Cloud-based monitoring of smart grids data can improve energy usage levels via energy monitoring and continuous feedback to users along with real-time pricing information of the billed amount. Smart grids help in dynamic pricing, for example, different tariff tiers during peak loads and off-peak loads. Condition monitoring data collected from power generation and transmission systems can help in detecting faults and predicting outages.

The next use of the IoT is in Energy Renewable Energy System design. Due to the variability in the output from renewable energy sources (such as solar and wind), integrating them into the grid can cause grid stability and reliability problems. The IoT-based systems integrated with the transformer at the point of interconnection measure the electrical variables and how much power is fed into the end to monitor and control over and under production. The IoT devices can help shift more generation on green energy systems from conventional energy wherever possible.

Health management systems can predict performance of machines of energy systems by analyzing the extent of deviation of a system from its normal operating profiles. In the system such as power grids, real-time information is collected using specialized electrical sensors called Phasor Measurement Units (PMU). Analyzing massive amounts of maintenance data collected from sensors using high-capacity cloud systems in energy systems and equipment can provide predictions for impending failures.

2.3.6 City Management

In big cities, finding a parking space in the crowded city can be time consuming and frustrating. Smart parking makes the search for a parking space easier and convenient for the driver. Based on a combination of IoT sensors at parking slots with GPS devices on, cars can detect the number of empty parking slots and send the information over the internet to the smart parking applications, which can be accessed by the drivers using their smartphones, tablets, and in-car navigation systems (Srikanth et al., 2009).

Smart lighting for roads can help in saving energy by automatically switching street lights on and off based on natural light availability. Smart lighting for roads allows lighting to be dynamically controlled and also adaptive to ambient conditions. These, when connected to the internet, can be controlled remotely to configure lighting schedules and lighting intensity. Custom lighting configurations can be set for different situations such as a foggy day, a festival, etc. (Jara et al., 2014).

Smart roads provides information on driving conditions, travel time estimates, and alerts in case of poor driving conditions, traffic congestion, and accidents. Such information can help in making the roads safer and help in reducing traffic jams. Information sensed from the roads or traffic signals using video scanning technologies can be communicated via the internet to cloud-based applications and social media and disseminated to the drivers who subscribe to such applications; for example, in a case study of a distributed and autonomous system of sensor network nodes for improving driving safety on public roads, the system can provide the driver and passengers with a consistent view of the road situation a few hundred meters ahead of them or a few dozen miles away, so that they can react to potential dangers early enough (IEEE, 2016).

Structural health monitoring is another area where the IoT network of sensors monitor the vibration levels in the structures such as bridges and buildings to analyze and assess the health of the structures. This helps detect cracks and mechanical breakdowns, locate the damages to a structure, and also calculate the remaining life of the structure and advance warnings can be given in the case of imminent failure of the structure (Shah & Ilyas, 2016). Previous literature has proposed an environmental-effect removal-based structural health monitoring scheme in the IoT environment (Nithya et al., 2017) where it has explored energy harvesting technologies of harvesting ambient energy, such as mechanical vibrations, sunlight, and wind.

Another application of the IoT can be in surveillance of infrastructure, public transport, and events in cities required to ensure safety and security. Citywide surveillance infrastructure comprising of a large number of distributed and internet-connected video surveillance cameras can be created. The video feeds from surveillance cameras can be aggregated in cloud-based scalable storage solutions. Cloud-based video analytics applications can be developed to search for patterns of specific events from the video feeds.

The IoT systems can be used for monitoring the critical infrastructure cities such as buildings, gas, and water pipelines, public transport, and power substations. Multimodal information such as sensor data, audio, and video feeds can be analyzed in near real-time to detect adverse events. This can help in rehabilitation or evacuation, for example, in the case of natural calamities to prevent loss of life.

2.3.7 Logistics

Currently, the biggest application of the IoT is in logistics fleet tracking. Vehicle fleet tracking systems use GPS technology to track the locations of the vehicles in real time. The vehicle locations and routers data can be aggregated and analyzed for detecting hurdles such as traffic congestions on routes, assignments and generation of alternative routes, and supply chain optimization. Earlier research has proposed a system that can analyze messages sent from the vehicles to identify unexpected incidents and discrepancies between actual and planned data, so that remedial actions can be taken (Moeinfar et al., 2012). An extension to this is shipment monitoring solutions for transportation systems that allow the monitoring of the conditions inside containers; for example, containers carrying fresh food produce can be monitored to prevent spoilage of food. The IoT-based shipment monitoring systems use sensors such as temperature, pressure, and humidity, for instance, to monitor the conditions inside the containers and send the data to the cloud, where it can be analyzed to detect food spoilage (Bahga & Madisetti, 2014). Previous research has proposed a cloud-based framework for real-time fresh food supply tracking and monitoring. Another earlier research deals with Container Integrity and Condition Monitoring using a Vibration Sensor Tag (Bukkapatnam et al., 2012) where a system was proposed that can monitor the vibration patterns of a container and its contents to reveal information related to its operating environment and integrity during transport, handling, and storage.

The IoT-based Remote Vehicle Diagnostics can detect faults in the vehicles or warn of impending faults. These diagnostic systems use on-board IoT devices for collecting data on vehicle operations such as speed, engine RPM, coolant temperature, fault code number, and status of various vehicle subsystems. Modern commercial vehicles support on-board diagnostic (OBD) standard such as OBD-II. OBD systems provide real-time data on the status of vehicle subsystems and diagnostic trouble codes, which allow rapid identification of the faults in the vehicle. The IoT-based vehicle diagnostic systems can send the vehicle data to centralized servers or the cloud, where it can be analyzed to generate alerts and suggest remedial actions.

2.3.8 Healthcare

Wearable IoT devices allow continuous monitoring of physiological parameters such as blood pressure, heart rate, and body temperature that can help in continuous health and fitness monitoring, for example, a FitBit wristband. It captures information on total steps traveled, calories burnt, etc. Remote telemedicine has allowed remote monitoring of devices in hospitals by doctors sitting in some other place. They can analyze the collected healthcare data to determine any health conditions or anomalies. Previous literature has proposed a mobility approach for body sensor network in healthcare—a WSN compatible with small sensors having features of integrated ECG, accelerometer, and SpO2.

2.3.9 Industrial

Industrial machines with built in PLC and data ports have long been able to capture information on the machine's working and operating parameters; however, they have always operated in silos and hence data retention has been poor. It has been difficult to correlate in a multimachine-operating environment for boosting efficiency and productivity. The IoT-ready industrial machines have now made information more digitally accessible for further analysis.

Machine prognosis refers to predicting the performance of a machine by analyzing the data on the current operating conditions and how much deviations exist from the normal operating condition. It can help determine the cause of a machine fault or monitor the operating conditions such as temperature and vibration levels. Sensor data measurements are done on timescales of a few milliseconds to a few seconds, which leads to the generation of a massive amount of data. Usually, machines start giving deviation in readings before an impending failure, which can now be correlated.

Air Quality Monitoring inside industrial facilities can be done using the IoT. Air pollution is caused by harmful and toxic gases such as carbon monoxide (CO), nitrogen monoxide (NO), nitrogen dioxide, etc., that happens due to emissions and by-products, which can cause serious health problem to the workers. The IoT-based gas monitoring systems can help in monitoring the indoor air quality using various gas sensors. The indoor air quality can be placed in different locations. WSN-based IoT devices can identify the hazardous zones, so that corrective measures can be taken to ensure proper ventilation (Chang & Guo, 2006) and previous literature has presented a hybrid sensor system for indoor air quality monitoring that contains both stationary sensors and mobile sensors. A previous study on indoor air quality monitoring using a WSN (Bhattacharyya & Bandhopadhyay, 2010) provided a wireless solution for indoor air quality monitoring that measures the environmental parameters such as temperature, humidity, gaseous pollutants, aerosol, and particulate matter to determine the indoor air quality.

2.3.10 Environmental Monitoring

The environment is a prime area of concern as global warming and other factors are causing irreversible damage.

The IoT applications for smart environments include:

- *Weather Monitoring*: Using the IoT-enabled weather balloons or weather monitoring stations powered by the IoT, it is possible to collect data from different sensors, such as temperature, humidity, and pressure, and to correlate to predict the climate. AirPi is a weather and air quality monitoring kit that can be used for recording and uploading information about humidity, temperature, light levels, air pressure, carbon monoxide, UV levels, nitrogen dioxide, and smoke levels to the internet-based cloud.

- *Air Pollution Monitoring*: The IoT-based air pollution monitoring system can monitor the emission of harmful gases by factories and automobiles using gaseous and meteorological sensors. The collected data can be analyzed to make informed decisions on pollution control approaches. This can be extended to water pollution as well as discharges from factories or even groundwater quality monitoring.

- *Noise Pollution Monitoring*: The IoT sensors can help in generating noise maps for cities, which can help the policy maker in making policies to control noise levels near residential areas, schools, and parks (Eldien, 2009).

- *Forest Fire Detection*: The IoT-based forest fire detection system uses a number of monitoring nodes deployed at different location in a forest. Each monitoring node collects measurements on ambient condition including temperature, humidity, light levels, etc. Early detection of forest fires can help in minimizing the damage (Liu et al., 2011). Earlier research has presented a forest fire detection system based on a WSN. The system uses multicriteria detection, which is implemented by the artificial neural network (ANN). The ANN fuses sensing data corresponding to multiple attributes of a forest fire such as temperature, humidity, infrared, and visible light to detect forest fires.

- *River Flood Detection*: The IoT-based river flood monitoring system uses a number of sensor nodes that monitor the water levels using ultrasonic sensors and flow rate using velocity sensors. This can help raise alerts when rapid increase in water level and flow rate is detected (Lee et al., 2008). Earlier research has described a river flood monitoring system that measures river and weather conditions through wireless sensor nodes equipped with different sensors (Chang & Guo, 2006). Another study has described a motes-based sensor network for river flood monitoring that includes a water level monitoring module, network video recorder module, and data processing module that provides floods information in the form of raw data, predict data, and a video feed.

- *Agriculture*: A smart irrigation system can improve crop yields while saving water. These use the IoT devices with soil moisture sensors to determine the amount of moisture on the soil and release the flow of the water through the irrigation pipes only when the moisture levels go below a predefined threshold. A live case example is Cultivar's Rain Could, a device for smart irrigation that uses water valves, soil sensors, and a Wi-Fi-enabled programmable computer. Smart Green House controls temperature, humidity, soil, moisture, light, and carbon dioxide levels that are monitored by sensors and climatology conditions controlled automatically using actuation devices (Akshay et al., 2012). A system can be provisioned that uses a WSN to monitor and control the agricultural parameters such as temperature and humidity in real time for better management and maintenance of agricultural production.

2.4 Conclusion

The IoT has widespread use in both consumer and industrial scenarios. The IoT has now expanded to IoE, which includes human beings and live forms, which are not objects in the stricter terms but need to be active actors in business use cases for the IoT to succeed.

It helps benchmark production lines and optimize efficiency and reliability, together with other practices like lean manufacturing or six sigma management models. Numerous organizations have adopted IIoT to streamline their manufacturing operations (Ramakrishnan & Gaur, 2016), many of which have been listed in the above sections. Manufacturing in food and beverages, chemical, pharmaceutical, consumer packaged goods, textiles, or biotechnology industries is becoming more profitable based on the power of the IoT. The IoT helps organizations differentiate their product offerings by bundling software and digital value adds with the physical products or traditional offerings. The IoT also helps in launching new revenue streams and business models such as pay for usage models, thereby allowing consumers to take control of assets without incurring capital costs. This has been observed in agriculture machinery provisioning on a rental basis or construction equipment. Companies are adopting the IoT both inside the enterprises for tracking of assets and outside by introducing IoT in their end products. As the cases have shown, companies have benefited from new revenue streams, higher assets productivity, and lower operational costs of servicing end products.

However, significant challenges still exist in terms of infrastructure constraints, understanding of the IoT technology, steep initial investment, skill availability, and concerns on feasibility and returns.

References

Aggarwal, R., Das, M. L. 2012. RFID Security in the Context of "Internet of Things." *First International Conference on Security of Internet of Things* (pp. 51–56). Kerala, doi: 10.1145/2490428.2490435.

Akshay C., Nitin K., Abhfeeth K. A., Rohan K., Tapas G., Ezhilarasi D., Sujan Y. 2012. Wireless sensing and control for precision Green house management. *2012 IEEE Sixth International Conference on Sensing Technology (ICST)*.

Bahga, A., Madisetti, V. 2014. *Internet of Things – A hands-on approach*, ISBN: 978-0996025515. Hands-On Approach Textbooks. http://www.hands-on-books-series.com/

Bhattacharyya, N., Bandhopadhyay, R. 2010. Electronic Nose and Electronic Tongue. *In Nondestructive evaluation of food quality*, pp. 73–100, Springer, Berlin, Germany.

Biddlecombe, E. 2009. UN Predicts "Internet of Things." BBC News. Retrieved July 6, 2009 from http://news.bbc.co.uk/2/hi/technology/4440334.stm

Bukkapatnam, S., Kamarthi, S., Huang, Q., Zeid, A., Komanduri, R. 2012. Nanomanufacturing systems: Opportunities for industrial engineers. *IIE Transactions*, 44(7), 492–495, doi: 10.1080/0740817X.2012.658315, Taylor & Francis.

Butler, D. 2010. Computing: Everything, everywhere. *Nature*, 440, 402–405. doi: 10.1038/440402a.

Chang, N., Guo, D.-H. 2006. Urban flash flood monitoring, mapping and forecasting via a tailored sensor network system. *Proceedings of the 2006 IEEE International Conference on Networking, Sensing and Control 2006*, 23–25, pp. 757–761, April 2006.

Dodson, S. 2008. The net shapes up to get physical. *Guardian.* https://www.theguardian.com/technology/2008/oct/16/internet-of-things-ipv6

Eldien, H. H. 2009. Noise Mapping in Urban Environments, Application at Suez City cente, International Conference on Computers & Industrial Engineering, (ICCIE) July 6–9, 2009, IEEE: Troyes, France.

Emil, J. 2005. A wireless body area network of intelligent motion sensors for computer assisted physical rehabilitation. *Journal of NeuroEngineering and Rehabilitation*, 2, 6.

Energieeffizienz in der Produktion. 2017. Untersuchung zum Handlungs- und Forschungsbedarf. Retrieved November 11, 2017 from https://www.fraunhofer.de/content/dam/zv/de/forschungsthemen/energie/Studie_Energieeffizienz-in-der-Produktion.pdf

Ferguson, G. T. 2002. Have Your Objects Call My Objects. *Harvard Business Review*, 1–7. Retrieved March 1, 2018 from https://hbr.org/2002/06/have-your-objects-call-my-objects.

Gartner. 2017. https://blogs.gartner.com/smarterwithgartner/files/2017/08/Emerging-Technology-Hype-Cycle-for-2017_Infographic_R6A.jpg

Gershenfeld, N., Krikorian, R., Cohen, D. 2004. The Internet of Things. *Scientific American*, 291, 76–81, doi: 10.1038/scientificamerican1004-76.

Gigli, M., Koo, S. 2011. Internet of Things, services and applications categorization. *Advances in Internet of Things*, 1(2), 27–31. doi: 10.4236/ait.2011.12004.

Graham, M., Haarstad, H. 2011. Transparency and development: Ethical consumption through web 2.0 and the Internet of Things. *Information Technologies & International Development*, 7(1), 1–11, Research Article, 7–14.

Gubbi, J., Buyya, R., Marusic, S., Palaniswami, M. 2013. Internet of Things: A vision, architectural elements, and future directions. *Future Generation*, 29, 1645–1660.

IEEE. 2016. The 1st IEEE PERCOM workshop on security privacy and trust in the internet of things (SPT-IOT). *In Conjunction with IEEE PERCOM 2016*, Sydney, Australia.

Jara, A. J., Miguel A. Z., Antonio F. S. 2014. Drug identification and interaction checker based on IoT to minimize adverse drug reactions and improve drug compliance. *Pers Ubiquit Comput*, 18, 5–17, London: Springer-Verlag.

Kosmatos, E. A., Tselikas, N. D., Boucouvalas, A. C. 2011. Integrating RFIDs and smart objects into a unified Internet of Things architecture. *Advances in Internet of Things: Scientific Research*, 1, 5–12. doi: 10.4236/ait.2011.11002.

Lee, J., Kim, J.-E., Kim, D., Chong, P. K., Kim, J., Jang, P. 2008. *RFMS: Real-time Flood Monitoring System with Wireless Sensor Networks*, pp. 527–528, doi: 10.1109/MAHSS.2008.4660069.

Li, B., Yu, J. 2001. Research and application on the smart home based on component technologies and Internet of Things. *Procedia Engineering*, 2087–2092. doi: 10.1016/j.proeng.2011.08.390.

Lianos, M. D. 2000. Dangerization and the end of deviance: The institutional environment. *British Journal of Criminology*, 40, 261–278.

Liu, Y., Xu, H., Teo, K. 2018. Forest fire monitoring, detection and decision making systems by wireless sensor network. 5482–5486. 10.1109/CCDC.2018.8408086.

Lombreglia, R. 2010. The Internet of Things. *Boston Globe*.

Moeinfar, D., Shamsi, H., Nafar, F. 2012. Design and Implementation of a Low-Power Active RFID for Container Tracking @ 2.4 GHz Frequency. *Advances in Internet of Things*, 2(2), 13–22, Scientific Research.

Nithya, R. Rajaduari, M. Ganesan, Ketan Anand. 2017. A survey on structural health monitoring based on internet of things. *International Journal of Pure and Applied Mathematics*, 117(19), 389–393. http://www.ijpam.eu.

Ramakrishnan, R., Gaur, L. 2016. *Application of Internet of Things (IoT) for Smart Process Manufacturing*. Springer, Information Systems Design and Intelligent Applications.

Razzak, F. 2012. Spamming the Internet of Things: A possibility and its probable solution. *Procedia Computer Science*, 10, 658–665, doi: 10.1016/j.procs.2012.06.084.

Sanjana, K. V., Yuen, C., Tushar, W., Li, W.-T., Wen, C.-K., Hu, K., Chen, C., Liu, X. 2016. System design of internet-of-things for residential smart grid. *IEEE Wireless Communications*. 23, 90–98, doi: 10.1109/MWC.2016.7721747.

Shah, S. H., Ilyas Y. 2016. A survey: Internet of things (IOT) technologies, applications and challenges. *2016 the 4th IEEE International Conference on Smart Energy Grid Engineering*, Peshawar, Pakistan.

Shao, W. A. 2009. Analysis of the development route of IoT in China. *Perking China Science and Technology Information*, 24, 330–331.

Srikanth, S. V., Pramod, P. J., Dileep, K. P., Tapas, S., Mahesh, P., Sarat, C. B. N. 2009. Design and implementation of a prototype smart PARKing (SPARK) system using wireless sensor networks. *2009 IEEE 23rd International Conference on Advanced Information Networking and Applications Workshops (WAINA)*, Bradford, pp. 401–406, doi: 10.1109/WAINA.2009.53.

United Nations. 2012. United Nations Conference on Sustainable Development, Rio+20. Retrieved from UN Sustainable Development Goals: https://sustainabledevelopment.un.org/rio20c, June 20, 2018.

Want, R. 2006. An introduction to RFID technology. *IEEE Pervasive Computing*, 5, 25–33.

3

Current Technologies

3.1 Introduction

As the Internet of Things (IoT) proliferates in almost every walk of our daily lives, be it through transportation, health services, or other areas, further potential will be realized in combination with related technologies and approaches like virtualization-based cloud computing, future internet of everything, big data, semantics, virtual sensors, robotics, artificial intelligence, virtual reality, low form factor devices, and less energy consuming next-generation network layers. Numerous industry analyses have shown that we are moving toward a web of connected sensors, services, and live entities such as people. The IoT with millions of connected technologies require scalable architecture, nontraditional compute power of several orders higher than transaction systems, transition to open systems with common data interchange formats, and knowledge-based applications that can utilize this volume of data. This comes with its own set of technological and conceptual challenges, autonomous objects intelligent enough to take critical decisions, governance issues, and man-machine seamless interfaces using smart object discovery, power management, and broadcast of presence and capabilities. The concern of uniquely identifying each device on the internet seems to be have been addressed with the migration to IPV6, but the concern of addressability still persists, which is the existence of a translation scheme for such a huge number of addresses. The focus in this chapter is on the IoT architectures, cloud computing aimed at providing the infrastructure for the IoT with collaborative computing, embedded platforms at device levels, message sharing protocols, data handling environments, governance framework for ensuring availability and security, and finally failure detection and recovery mechanisms.

To ensure things or devices are available without any time or place limitations, hence ubiquitous computing is a necessity, which can be ensured using a number of device communication protocols which include wireless sensor networks (WSN), Bluetooth Low Energy (BLE), and LT-IoT. Also, devices and infrastructure need to support the unique IoT requirements of scalability, integration, volume processing, and real-time analytics, for which a range of new technologies like cloud, fog, and mobile computing has evolved. In parallel, software defined networks have been developed, and containers for embedded computing. Cloud computing extends delivery of hardware and resources using the internet-based on-demand and almost unlimited scalability potential. These reduce the ownership costs while ensuring on-demand computing in terms of scale up or scale down capacity by way of resource pooling and rapid elasticity and pay per use models.

To understand what all current technologies and future enhancements will contribute significantly to the IoT, we need to understand the fundamental building blocks of this stack. There are six main requirements which need to be addressed for the IoT to be functional as discussed below.

3.1.1 Identification

Objects have been traditionally identified using uCode (ubiquitous code) and electronic product codes (EPC). Identification differs from addressability in the sense that while a unique address will be needed to communicate on the internet, these addresses can be recycled, hence a unique identifier is required in addition to the address to ensure that the object initiating or receiving data packets is authorized to do so. For this reason, current desktops have a unique machine identification (MACID) while mobile phones have a unique IMEI number.

3.1.2 Sensing

This refers to gathering digital readings from an object and sending it to the network for further storage. Single Board Computers (SBCs), which now come with multiple sensors such as Temperature Sensor (such as DHT11 and DHT22), IR sensors, Voice sensors, Gas Sensors, and consist of inbuilt network cards and Wi-Fi ports that support TCP/IP and enforce security functionalities are used to create IoT products.

3.1.3 Communication

The IoT-based technologies for communication connect different smart objects together for services delivery based on data. Hence the IoT nodes should operate with minimal power requirements since it is pervasive and almost always connected with sensors for real-time data capture. Examples of communication protocols that are commonly used in the case of the IoT are Wi-Fi, Bluetooth, LTE-Advanced, IEEE 802.15.4, and Z-wave and to a limited distance range Near Field Communication (NFC), RFID, and ultra-wide bandwidth (UWB).

3.1.4 Computation

Small form factor boards like the Arduino or RaspberryPi require minimal compute instructions, and hence special operating systems known as real-time operating systems. These are smaller in size, can load into with minimal memory requirements, and run on processing units such as SOC's, microcontrollers, and microprocessors, to provide computational power to IoT devices.

3.1.5 Services

The IoT devices should be able to publish services that can be discovered by other interested objects. These services can help in identification or detection, record information, create collaborative platforms, or are pervasive. Examples of information aggregation category include smart healthcare and smart grids while collaborative-aware category examples are smart transportation integrated road, rail, waterways, and air services.

3.1.6 Semantics

This refers to the ability to provide required services using data recognition and analysis using different machines. Although this area is still grey for the IoT, in the World Wide Web this is much more defined using semantic web such as web ontology frameworks. Similarly, the EXI–Efficient XML program also provides semantics for the web.

3.2 Cloud Deployment Models

Cloud infrastructure can be provisioned in terms of service layer into Software as a Service (SaaS), Platform as a Service (PaaS), and infrastructure as a service (IaaS) models. Each has its own merits and demerits. While SaaS only provides the software stack to provision cloud enabled applications, IaaS offers an entire set of hardware needed to build a customized computing environment. PaaS, in between, provides a development environment along with remote programming model support. In terms of deployment models, cloud can be set up as a private cloud, public cloud, or a hybrid cloud depending on whether the environment is exclusively reserved for an organization or shared across many users. With the convergence of IoT and cloud being a reality, streaming data in a high-performance way using compression or packet optimization is becoming a priority for the IoT applications.

3.3 Relevant Technologies

There are many other areas that are converging to support and enable IoT applications, which include the defined IoT architectures, communications stack, identification algorithms and protocols, network discovery protocols, next-generation network technology, software based algorithms, hardware technologies, data and signal processing techniques, search engines, network management, power and energy, security and privacy, interoperability, and standardization.

3.4 Small-Form Factor Computing

In the world of small-form factor computing, data acquisition services have been traditionally supplied via add-on I/O modules in one stackable I/O form factor or another. While this has served the embedded market well and will continue to usher in many applications, the continuing requirements for smaller, lighter, less power consuming and low-price points are enabling rapid changes not only in embedded single-board computers, but also in the way data acquisition is implemented, for example, Raspberry PI. The Raspberry PI supports multiple operating systems ranging from Windows, Android to Raspberry native OS, and even a RISC-based (Reduced Instruction Set) OS. It has built in USB ports, a Bluetooth transreceiver, infrared transmitters, a Wi-Fi port, and supports an SD card for storage. Every year, we find new proprietary form factors being released in many cases by integrated solution providers who make their boards unique to their enclosures. Currently, available solutions in the markets are:

- Raspberry PI
- Intel Edison and Galileo
- Arduino Uno and Mini

3.5 Web Services

Web services refer to a software program published by one device that can be read using other software programs knowns as a web service client by another device. They exist in two types, REST or SOAP, each of which varies in structure and content. Web services can be used for machine to machine (M2M) transmission and provides an object-oriented web interface to data stored in underlying database objects or as a mashup to aggregate web services from different sources providing a single window view of information.

As per the W3C (2004) definition: "web service is a software system for supporting inter-operable machine to machine interaction over a network."

3.6 Wireless Sensor Networks (WSN)

For smart objects to communicate in the IoT environment, a distributed wireless sensor network (WSN) is required to address multiple challenges of detecting relevant devices, transferring small volumes of data at repetitive intervals, handle error correction in case of breakages, and ensure uptime of transmission. New initiatives like the DARPA SENSIT and NSF programs are working on further refining these networks. These networks follow the pattern of traditional networks and can exist in star, ring, bus, and tree and mesh formats. The challenge is to increase the power lifetimes of these devices by more effective energy management during transmission, and designing microelectromechanical systems, which can use renewable energy sources such as solar, vibration, or thermal. RFID tags have transponder microcircuits, which have an LC tank circuit that can store power received from incoming signals and use it to send responses. TDMA technology becomes more effective since it can sleep and wake up during transmission intervals.

3.7 Virtual Sensors

Sensors are small devices that can report about the environment such as temperature, humidity, and moisture by monitoring events, changing state, and then generating output (Soloman, 1999). A virtual sensor is an abstract form of a physical sensor that can detect events produced by a physical sensor and takes two kinds on inputs one from the physical sensor and the other from the user (Bose et al., 2015). While physical sensors have a fixed behavior, a virtual sensor acts on a need basis. While for a physical sensor the information source is always direct, for a virtual sensor it may be indirect as well. Physical sensors are related to a specific domain such as chemical, thermal, magnetic, electrical, and mechanical and hence is bounded. Virtual sensors can span across multiple domains such as healthcare monitoring systems (Liu et al., 2009). Cloud-based virtual sensors can be invoked remotely such as climate detecting sensors published as a software service for consumption by interested entities. Virtual sensors also help address memory and power consumption challenges and also act based on defined threshold values, classifying it as an event when the value is crossed. Also, unlike physical sensors which are affected by transformation of physical state with each event, thereby making next readings more error-prone, virtual sensors do not have these constraints (Evensen & Meling, 2009). The following taxonomies virtual sensors are considered important in a PaaS or SaaS environment in the cloud.

- *Singular*: a single physical sensor is coshared between multiple applications.

- *Aggregator*: Multiple physical sensors are combined to provide consolidate data.
- *Accumulator*: A single virtual sensor mapped to many physical sensors which can be accessed separately or combined.
- *Selector*: With limited options sets values to pick up from by the sensor.
- *Qualifier*: With defined quality of service for priority processing of data.
- *Predict and compute*: allows prediction of forecast based on current datasets.

3.8 Fog Computing

This refers to an architecture in the IoT where the compute power is hierarchically distributed from the edge of the network to the core (Bonomi et al., 2012). Some applications that have low latency requirements cannot benefit from the adoption of the cloud as the time taken for data and response transmission over the cloud can be significantly high (Pao & Johnson, 2009). Hence, data needs to be stored in the vicinity of the device and also meet additional requirements of mobility support, geo identification, and location awareness. This is achieved by extending the cloud to the edge of the network also referred to as fog computing. Hence, fog was designed for applications that cannot operate with the cloud but have the same set of requirements namely virtualization, multitenancy architecture, shared network, compute, and storage. The following are situations where fog computing can be used effectively (Cristea et al., 2013).

- Applications with low latency such as gaming or media rich ones.
- Applications with wide geo spread such as oil and gas pipelines being monitored.
- Mobile applications such as smart transport systems.
- Distributed computing requirement applications such as intelligent power grid and distribution systems.

Hence, data sources become numerous and aggregating only relevant data and not real data in this scenario. Hence, while today each device may have a user, in the future we are bound to see more and more devices grouped as systems and more autonomous in nature. The use case of fog computing is the envisaged urban traffic lighting system to be paired with self-drive vehicles. The use case involves creating multiple signals with sensors that can sense vehicles coming and adjust signal duration and timings accordingly. The objectives are to prevent accidental collisions, ensure smooth traffic movements, and collect and analyze vehicular movement data. While the first objective requires real-time processing of data, the second is near real-time, and the last is more of a batch processing nature. Hence, the first part requires use of fog computing in the order of <10 ms.

The requirements that are met are as follows:

- Low latency with millisecond response time for accident prevention
- Geo distributed middleware with the ability to push rules and policies to individual light systems
- Networking with extremely rugged systems with interfaces
- Cloud integration for sending data for analytics related to trends and patterns of vehicles, collisions and movements, and environmental pollutants
- Consistency between different nodes
- Multitenancy and integrated hardware, software, and application
- Multiple providers and stakeholders

3.9 Artificial Intelligence (AI)

Here, AI refers to intelligence demonstrated by machines. However, as things become more mainstream, they get removed from the purview of the AI such as optical character recognition is no longer considered part of AI. Recent advancements have led to the evolution of self-drive cars and drones and deep learning is further helping evolve trends from machine datasets. These datasets may be structured like text and databases or unstructured like images and videos and have the potential to be an important source of knowledge from the mid-twentieth century (McCarthy, 1959; McCarthy, 1977). With statistical methods and libraries, there have been significant developments of applications (Weikum et al., 2009). The objective of AI is to analyze the situation and take actions with a probability of increasing its goal of successful outcomes. These goals can be either predefined or goals can be derived from training data such as in the K-Nearest Neighbor algorithm. Some AI algorithms are self-learning from data and the objectives are self-planning and learning through automation. The next generation of AI based on natural language processing confers on machines the ability to read and write in human language (Russell & Norvig, 2009). The concept uses, other than word matching, lexical meaning matching to group data and categorizing searches. Machine perception or computer vision is yet another category that can use sensor inputs such as camera images, voice, and video and it is used for face recognition or speech recognition. AI has predominant usage in robotics, and assists in movement based on experience and can navigate shop floors with different layouts. AI can work on top of the IoT generated machine data to make machines more autonomous and there are numerous examples integrating both of these technologies. iRobot Roomba, a vacuum cleaner, has AI to map home locations and adapt to different surfaces or new items. Apple Siri is an AI embedded inside an internet-connected mobile phone device. Nest Labs thermostat solutions, which are IoT devices, not only captures room temperature and control but also uses AI to learn about user preferences and accordingly suggest them. Tesla Motor electrical cars operates as a fleet network and all the cars share collective learnings as part of AI. This consists of connected sensors to generate raw data for further analysis.

3.10 Virtual Reality

This refers to an interactive cyber and auditory and visual sensory experience. In a simulation, this can help users work virtually in an environment not accessible normally, such as engine chambers, high heat and pressure environments, or medical procedures and surgery training in a human body. These are useful in telepresence and telerobotics, although they come with numerous challenges such as health and safety and privacy (Rosenberg, 1992). Virtual reality (VR) and the IoT share the same objective of merging the physical and cyber spaces, although the approach is different as VR makes the digital world real while the IoT makes physical objects digitally controllable. The combination of VR and the IoT is used in telepresence, and two products, Empathy VR and OdenVR Telepresence Robot, pair a VR head onto a mobile robot to give a very realistic presence. Another company, Eye Create Worlds, has worked on combining the IoT sensors through a city, which can be used by an operator in a virtual world to monitor and optimize rail networks.

3.11 Big Data

Big data is an acronym to describe huge volumes of structured or unstructured data generated by devices and machines that cannot be handled by traditional database systems and topology due to the compute power requirements and the incremental data volumes (Diebold, 2012). Usually, big data, unlike transactional data, does not change with time and only new data keeps getting added. Notable traditional sources of this could be social media platforms like Facebook and Twitter that generate millions of data packets every minute, clicks of millions of pages of high-volume websites, or more recently data getting generated by sensors in manufacturing plants, telecom systems, or transportation systems. Big data can be used to provide insights and predict trends or patterns based on historical data points. Existing big data frameworks such as Apache Hadoop need to be augmented to effectively store, manage, and extract value from continuous streams of sensor data and organization's also need to understand the significance of maintaining data (Lynch, 2008). For instance, it is estimated that connected cars will send 25 gigabytes of data to the cloud every hour. The biggest challenge will be to make sense of this data, identifying data that can be consumed and quickly acted upon to derive actionable events.

The key considerations in analytics of big data include the following:

- Cost of storage and processing of huge data
- Pressure to gain insight into data quickly
- Diversity of data formats
- Combining insights across formats, volume, and speed of processing
- The fast-changing technology landscape of big data analytics.

To work with big data for useful insights, organizations typically adopt a three-step approach. The steps in this approach have been listed below:

- Understanding the value proposition of big data
- Evolve a big data strategy covering:
  - Acquisition
  - Storage and processing
  - Analysis and action
  - Management
- Map out the solution for analytics
  - Setting up a big data project
  - Choice of a big data analytics solution

3.12 Blockchain and Cryptography

Cryptography refers to encrypting and decrypting information using mathematical models and string keys to ensure data is visible only to the sender and authorized receiver. Broadly used, cipher algorithms include AES, 3DES, and DES. A blockchain refers to a series of transactional information stored chronologically in the internet with copies referable to legitimate users. In blockchain technology, cryptography is used for ensuring past records cannot be tampered with and also securing the identity of the sender of information. Information stored in a blockchain can be of any types— transfer of money, agreements, documents, messages, etc., and requires confirmation from multiple devices in the network and once stored it cannot be tampered without permission and knowledge of the entire community. This information is stored in multiple copies in different locations, hence ensuring redundancy and failsafe records. The convergence of the IoT and blockchain is a happening scenario today, since the latter can provide a solution to the privacy and security concerns of the former. The IoT data that are critical can be distributed over multiple servers ensuring no single point of failure. Qtum, a blockchain company, is developing the IoT platform that connects Ethereum smart contracts with Bitcoin's prolific blockchain. As per IDC, by 2019, 20% of all IoT deployments will have blockchain services enabled and will determine how devices will communicate with each other. The benefits of combining these two technologies will include building trust between devices, eliminating tampering risks, reducing costs through direct interaction and avoiding intermediaries, and reducing settlement time leading to higher speed. In the case of IIoT, business models require a number of contracts between contracting parties such as in a supply chain or shared production models. A real scenario use case is in the insurance sector where the IoT devices can be used to assess claims and record data, while blockchain can be used to store the data for autonomous claim settlement.

3.13 Semantic Technologies

This refers to separating data and content files from the application code so that machines and people can reason and understand this in real time. For the IoT data to transition from being raw data to insights, data with semantic annotations, interoperable formats and adaptable and context-aware solutions are required. Although the IoT has defined frameworks, semantic technologies are needed for integration of IoT data. XML is one of the forefront semantic languages currently being used for M2M communication; however, it lacks a semantic model and has a surface model only and there are multiple domain specific tags that cannot be machine interpreted unless programmed. As the IoT devices tend to become more autonomous and plug and play, there is a need for extending the semantic web to a semantic sensor web to give information a well-defined meaning for autonomous interaction. Resource Description Framework (RDF), a W3C standard, consists of triplets or sentences with subject, property, and object, for example, "Sensor," "hasType," "temperature" or "Node," "hasLocation,"

"RoomX." Every resource that can be an IoT device will have a universal identifier or URI. Semantic technologies can make the IoT more interoperable, enable data access, help in resource discovery by ways of broadcast, and help in processing of data.

3.14 Fuzzy Logic

In some IoT situations, setting values based on traditional binary state data may not be practical since it depends on numerous other variables, for example, the case of a home monitoring system where the optimum temperature may be dependent on variables like external environment and pressure and humidity. Fuzzy logic with its capability of evaluating multiple states simultaneously can help simulate human intelligence in cases where the inputs may change suddenly and often unpredictably (Nikravesh & Zadeh, 2007). Using words for computation may significantly improve the decision-making capacities of the computer and hence this will be the future direction for the IoT sensor data, which will transition from the current numerical on-off state analog data to a more fuzzy data. This is also useful where we have limited or incomplete data inputs or want to analyze data very fast, for example, seeing an MIS report spend limited time checking the subtotals to see if the data is accurate as per our earlier observations ("using common sense"). Hence, fuzzy logic data, which is more qualitative and represents degrees of state between 0 and 1, will be captured and pushed to the server and this fusion with the IoT can give different benefits. Cases could be smart cities adjusting the temperature and energy consumption based on outside environmental conditions, managing a product life cycle based on degree of acceptance by a target user, or by the level of product maturity when benchmarked against existing products. Areas in healthcare that use IoT can also benefit by the application of fuzzy logic as was observed in a study of diabetes control (Grant, 2007).

3.15 Next Generation Networks (NGN)

Telecom networks such as 4G/5G will form a major component of provisioning the IoT services; hence, operators are focusing on the NGN based on fiber-optic cables in which all information and services will be transferred as one IP module and will be entirely packet based thereby enabling Quality-of-Service (QoS) facilities. The core network will shift from current Packet Switching (PSTN) to an IP based (VoIP) transport layer replacing technologies like X.25

and frame relays with IP VPN. The wired network will see a change from current combination of voice and xDSL to DSLAM (digital subscriber line access multiplexer), hence, making voice switching equipment redundant.

3.16 Future Internet

This refers broadly to the new architecture related research for the internet, to overcome the current limitations in terms of security and performance, required with the IoT connecting billions of devices. The approach consists of both technical, for example, differentiated services (DiffServ mechanism with low latency), reliable server pooling (RSerPool) and stream control transmission protocol; and nontechnical aspects such as social and governance that are being worked out by different researchers like The Global Environment for Network Innovations (GENI).

3.17 Nano-Networks

Small devices in the scale of nanometers or micrometers interconnected with each other with the ability to perform basic level computing, storage, or other simple tasks, are now getting used to expand capabilities for the IoT networks. These are dependent on either electromagnetic or molecular or communication. The former being based on transmission of electromagnetic signals using modulation and demodulation techniques on carbon nanotubes while the latter is based on molecule propagation using predefined paths or molecular motors (walkway based) or flow based in liquids or diffusion based in the case of gases.

3.18 Single-Point Sensing

As the IoT devices start getting placed everywhere for monitoring purposes, with reducing the cost of devices, a huge impact on power and security and management issues for each of these installed devices will start to occur. There is proposed a single-point sensing approach using mediation by way of infrastructure. These involve a network of the IoT sensors with higher telemetry and resolution connecting to the cloud for analytics; hence, using one single device stationed near the power box. Examples of these are microphones to detect breakages from a central location rather than having multiple microphones. Washington Research Foundation experimented with strategic placement of sensors to detect electric signals based on infrastructure-mediated sensing (Emil, 2005). Hence, strategic placement of resources and using high resolution recordings from infrastructure can be optimized using neural networks algorithms such as the AI toolchains of Google. This infrastructure-mediated single-point sensing is also being extended to industrial applications such as telecom network breakage detection.

3.19 Future Advancements

Privacy is an important consideration and new frameworks have/are being evolved for a low-cost solution to this problem in IoT projects, covering end to end from the point of data capture to data processing facilities assuming that entities are connected and communicating. Communication can be direct, that is, connected over a network at each point in time or indirect, that is, based on a service-based push or pull. Traditionally, the end points used to transfer data to a service platform and applications used to pull data from the platform but there were no end-to-end encryption mechanism. There has to be an end-to-end encryption between the endpoint and the application consuming the data. These are areas of security where we expect to see new developments in future with minimal use of computation power and energy for encryption algorithms.

3.20 Conclusion

The IoT being a recent development, many of its core and supporting technologies are still in a stage of evolution (Figure 3.1). In some areas, there

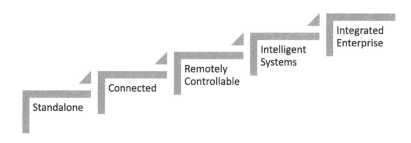

FIGURE 3.1
Transition from M2M to the IoT.

has been a total transformation as new technology advancement happens such as type of network connectivity while some of them are mostly stable such as the real-time low-footprint operating systems. The technologies are fraught with challenges, which may range from security to adoption and business feasibility. There are many supporting technologies that further contribute to the ecosystem complexity as a whole and require meticulous planning and execution to derive full benefits of automation and the IoT adoption. This chapter has tried to cover the building block technologies for the IoT and also touched upon the latest trends in different areas that have evolved to meet the volume, compute, and data transmission requirements of the IoT in a pervasive and connected world.

References

Bonomi, F., Milito, R., Zhu, J., Addepalli, S. 2012. Fog computing and its role in the internet of things. *Proceedings of the 1st Edn. of the MCC Workshop on Mobile cloud Computing*. Australia.

Bose, S., Gupta, A., Adhikary, S., Mukherjee, N. 2015. Towards a sensorcloud infrastructure with sensor virtualization. *Proceedings of the Second Workshop on Mobile Sensing, Computing and Communication,ser. MSCC '15* (pp. 25–30). NY, USA: ACM. doi: 10.1145/2757743.2757748.

Cristea, V., Dobre, C., Pop, F. 2013. Context-aware environ internet of things. *Internet of Things and. Inter-Cooperative Computational Technologies for Collective Intelligence Studies in Computational Intelligence*, 460, 25.

Diebold, F. X. 2012, September 21. *Development of the Term "Big Data."* PA: Penn Institute for Economic Research. doi: 10.2139/ssrn.2152421

Emil, J. 2005. A wireless body area network of intelligent motion sensors for computer assisted physical rehabilitation. *Journal of NeuroEngineering and Rehabilitation*, 2, 6.

Evensen, P., Meling, H. 2009. Sensor virtualization with self-configuration and flexible interactions. *Proceedings of the 3rd ACM International Workshop on Context-Awareness for Self-Managing Systems* (pp. 31–38). NY, USA: ACM. doi: 10.1145/1538864.1538870.

Grant, P. 2007. A new approach to diabetic control: Fuzzy logic and insulin pump technology. *Medical Engineering & Physics*, 29(7), 824–827, ISSN 1350-4533, https://doi.org/10.1016/j.medengphy.2006.08.014.

Liu, L., Kuo, S. M., Zhou, M. C. 2009. Virtual sensing techniques and their applications. *International Conference on Networking, Sensing and Control* (pp. 31–36), ICNSC '09.

Lynch, C. 2008. How do your data grow? *Nature*, 455, 28–29. doi: 10.1038/455028a.

McCarthy, J. 1959. Programs with common sense. *Proceedings of the Teddington Conference on the Mechanization of Thought Processes* (pp. 75–91). London, UK: Her Majesty's Stationary Office.

McCarthy, J. 1977. Epistemological problems of artificial intelligence. *Proceedings of the 5th International Joint Conference on Artificial Intelligence* (pp. 1038–1044). Cambridge: Mass.

Nikravesh, M., Zadeh, L. A. 2007. *Studies in Fuzziness and Soft Computing*, 217, Berlin, Heidelberg: Springer-Verlag.

Pao, L. Y., Johnson, K. E. 2009. A tutorial on the dynamics and control of wind turbines and wind farms. *American Control Conference*. NY.

Rosenberg, L. 1992. The Use of Virtual Fixtures As Perceptual Overlays to Enhance Operator Performance in Remote Environments. *Technical Report AL-TR-0089, USAF Armstrong Laboratory, Wright-Patterson AFB OH.*

Russell, S. J., Norvig, P. 2009. *Artificial Intelligence: A Modern Approach*, 3rd edn. New Jersey: Prentice Hall.

Soloman, S. 1999. *Sensors Handbook*. Online: McGraw-Hill handbooks. https://books.google.co.in/books?id=VvQRkAzT1h4C

W3C. 2004. Web Services Glossary § Web Service. February 11, 2004. Retrieved January 24, 2017, https://www.w3.org/TR/ws-gloss/

Weikum, G., Kasneci, G., Ramanath, M., Suchanek, F. 2009. Database and information-retrieval methods for knowledge discovery. *Communications of the ACM* (pp. 56–65).

4

Business Models

4.1 Introduction

The growth of the internet over the past two decades will be dwarfed as the "connected things" that are around us start going online and start collaborating (Gershenfeld et al., 2004). This chapter provides an overview of the IoT-enabled development and implementation of business models for the nondigital companies including traditional manufacturing or services companies. The initial section explores the role of digitization on changes in traditional business models and its growing significance and adoption over the last two decades. The other section deals with digital business model patterns that are becoming incorporated in physical industries, leading to a rapid merger of the cyber and the physical world. The IoT is the disruptive technology causing this transformation, which makes hybrid solutions possible that merge physical products and digital services. The different challenges faced while implementing innovative business models are also discussed in detail in implementing these IoT business models and solutions. Traditional business models may soon become outdated with the advent of connected objects since there will be a shift from selling digital features of an object rather than its physical attributes. These will give rise to a new generation of business models that have not been envisaged so far, bringing in better value proposition to both customers and suppliers. Traditional business models used to be centric to a firm; however, with the IoT mandating an integrated ecosystem model involving not only immediate customers and suppliers but also competitors and regulators, new business models will be collaborative and agile to swiftly respond to changes in a high technology-driven industry. Also, with digitization possible, changes can be swiftly rolled out as against the traditional product development life cycle. We will try to throw some insights on business models that can be introduced by traditional firms by focusing on customer needs. This technology can be used in a variety of business domains such as healthcare (Emil, 2005), manufacturing, transportation, e-commerce, retailing, and financial services since it has the potential to offer contextual information while being able to recognize the subject and their location in real time and an accurate manner. The digital world has extended our capabilities to abstract and simulate more than what could be done in the physical world context. Firstly, the ability to measure things—be it environmental impact, movement, surrounding environment, or scalars and vectors like speed, distance, and velocity comes with zero cost as sensors like accelerometers, temperature detectors, velocity measure, and luminosity detectors are able to measure and transmit information in a finely grained manner to central processing units. The second benefit is by enabling normal physical things to become "smart objects," which are not only passive transmitters but can also actively receive and act on instruction sets such as a valve shutting down dynamically based on pressure at different times of the day or depending on climate is more useful than a physical valve that can only act in a fixed manner. Thus, a high-resolution management also gets extended to the physical world. This has made digitization more significant to traditional companies that used to focus only on physical attribute sales. Current applications in this area that have been adopted as successful business models include compliance monitoring, preventive maintenance, remote diagnostics and service delivery, asset location tracking and contextual applications, and automatic provisioning and fulfilment.

4.2 Business Model Framework

Business models are used by a firm for generating revenue and ensuring profitability and must answer the questions "Who" is the target customer, "What" is our value proposition, "How" will this value be transferred and used, and "Why" justify the economic model behind it (Gassmann & Enkel, 2004). The business patterns determined are either "Add-on services based" or "White Label based," where an example of an add-on service is SAP, which is a German-based software company, and a white label example is Foxconn, a Taiwanese multinational electronics company. Previous literature has demonstrated some of the key issues when designing the IoT business models (Bucherer et al., 2012; Westerlund, 2014) due to diversity of devices, technological evolution, and immaturity and a heterogeneous ecosystem. Value creation using the IoT happens at the manufacturing layer, services or support layer, and as a cocreative partner powered by cognitive computing and intelligent self-thinking. However, a value-creating business model today may not be lucrative in tomorrow's environment (Osterwalder, 2004). The following are three distinct roles that the IoT can perform when it comes to business models.

1. It may be the a "value-adder" of the business model such as digital commerce where the product itself is sold due to its digital nature, examples can be Tesla, where a consumer can pay for a temporary raise in engine horse power and capacity.

2. It may be the primary constituent of the business model such as digital content dispensing machines that play video or audio on demand through set top boxes, or cell phones, such as Apple Music.

3. Loosely influencing models such as the IoT device-driven advertisements or billing systems that may not be directly relevant but which help with business objectives attainment.

There have been three technology waves in the internet evolution, namely Web 1.0 (web as a business infrastructure), Web 2.0 (web as social media), and Web 3.0 (Internet of Things). While in the first one, value was added by the data flow possible and it gave rise to business models like freemium, e-commerce, open source, etc., the second one added value by connecting users and their interaction, giving rise to business models like crowdsourcing and crowdfunding. The Web 3.0 is when objects and sensors add value and this will result in business models using digital content and services to get revenue rather than physical models. The previous evolution of technology-driven business models has shown that IoT will also have two implications: firstly, it will integrate customers and organizations thereby delegating self-service to the former (models such as e-commerce, user designed, and open source); secondly, it will extend repeat sale potential through after sales service by continuous stream of engagement (models such as buy, subscription, razor, and blade), and finally it will use analytics on user data to properly price a product or service offering. These models have a higher probability of adoption by digital companies where IT plays different roles based on cases and is by definition constitutive to the business itself. Business models such as freemium are applicable strictly to digital firms.

4.3 The IoT and Economics Context

The IoT can help build high-resolution control circuits, such as in the case of Google where it is possible to have tailor-made advertising to each individual browser and hence user. A high-resolution profile of the user can be drawn up in terms of location, browsing habits, and areas of interest, which is also lucrative to the advertisers. This is compared to the physical world like television or newsprint where mass marketing to all intended and unintended users is the only way possible. With the IoT, each object further becomes an advertising channel as, hypothetically, a wearable fitness band can even suggest food supplements based on activity and metered readings of body function. Hence, while IT was able to use only desktops and, of late, mobiles as advertising mediums, soon with the IoT wearables, goggles, clothes, and white goods will become mediums, thereby ensuring higher reach and more accurate target audience identification.

FIGURE 4.1
The IoT value provider chain.

Many companies have capitalized on the business value of the IoT and are slowly moving toward it. This extends much beyond operations cost savings. One aspect is of measurements on a real-time basis using sensors along with data transmission, and analysis of huge volumes of data using computational algorithms predicting trends, associations, and patterns relating to human behavior, buying trends, or consumption. This, collectively called big data, further requires a cloud-based scalable infrastructure. Many companies like Google and Microsoft are working on these accumulated data to provide machine-learning and deep-learning insights. This will help in transforming products for better adoption and customization based on user input data. Along with this is also an apprehension that the IoT may just be a hype and whether it really adds value. The value will come in diverse industries based on diverse needs, which are not fulfilled by available technology. Where the IoT holds promise, for example, is in timely delivery needs in logistics, remote healthcare facility development in medical science, needs for differentiation and revenue in products such as automobiles or white goods, and needs for remaining connected to customers after sales in heavy industries (Figure 4.1).

4.4 The IoT Business Models

As there is a change in view of the IoT solution as not only a technology but a business solution and shift in focus from a firm's business model to the connected system ecosystem model (Westerlund, 2014), new application classes that are putting together the advancements in the cyber and physical world are powering new startups and traditional brick and mortar enterprises by working on data, sensors, devices, and users (Thestrup et al., 2006). People are moving toward everything "smart," smart world, smart cars, smart transportation, and even smart healthcare (Smith, 2012). These applications will lead to automation services that may transition business and economy. Business transactions can happen in terms of:

- Physical product: procurement, order processing, and production and distribution (Alt & Zbornik, 2002)
- Information in terms of order, and product data sharing
- Money stream based on product prices (Andrejovska, 2011)

Business models have been defined by numerous authors (Timmers, 1998) and limitations of definition also stated (Morris, 2005) with some focusing on the business value it adds, the change it brings about, and role of participants (Negelmann, 2001). Many frameworks have been defined that have been adapted for creating new IoT business models (Osterwalder et al., 2015). There are challenges in the IoT of integration of multiple business operating in collaboration and many new projects have still not been focused in their approach. While digitization will definitely yield benefits with connected objects, the challenge in business models for the IoT is in how they prove that the combination of product and services is adding value, which has been the major cause of failures (Gordijn, 2002). Business models need to clearly define what is being offered to customers, identify the customers, and how profitability is ensured and thereby help identify business stream and cost models. In the next section, we shall discuss cases where firms have either adopted a value model that focuses on value creation, or the structural process model that focuses on how business operations are being performed. The common IoT architecture consists of the perception (sensors capturing data), network (transmission of captured data), and application layers (data processing with business intelligence) (Sun et al., 2012). Business models encompassing all three layers are extremely scarce but the ones that are successful have used all three in a novel and new model (Wu et al., 2010). The first business model is the MOP (multiple open platform) consisting of dimensions of technology, industry, policy, and strategy (Li & Xu, 2013). Also exists is the DNA (design, need, and aspirations) model (Sun et al., 2012) illustrating the relation between the three used in smart logistics, the value net model used in telecommunications, which focuses on customer and information sharing (Li & Xu, 2013), the business model canvas, which focusses on information usage for value creation, and the 2×2 matric dimension model used in the automobile industry, which uses a strategy of customer centric information sharing along with resource integration. With this background literature, what emerges is that business models broadly consist of the following dimensions: infrastructure of partner's resources and activities, value proposition, customer segments and channels, and financials including cost and revenue.

4.5 Case Studies

The following three cases are picked up from different sectors, namely home automation (Google Nest), industrial automation (one from GE and one from a global packaging company), and transportation services (OLA, India). While the home automation service is based on subscription revenue, industrial automation has two cases: one where the IoT enable customer information sharing and another where the assets are integrated to bring operational efficiency. Technology platforms form the core of business strategies (Cusumano & Gawer, 2002); however, with the IoT, there are barriers to monetizing such as identifying horizontal needs and also the managerial competencies required in this revised situation (Wurster, 2014). Efforts to prepare an IoT business ecosystem (Mazhelis, 2012) have been promising but today's solution does not scale up to the level that is firm focused without providing value to all participants.

The first case deals with Nest, which is an artificially learning controller for temperature, placed in homes, and which can store user preferences or adjust as per external and internal climatic environment. As users adjust the settings, the device becomes intelligent enough to predict and set heating and cooling for homes. The device is controlled by a mobile application and can be monitored or controlled remotely over the internet. The benefits to the customer include energy saving, cost efficiency, and convenience. It can also integrate with other digital systems to become a centralized controller, such as digital door locks and other connected appliances including electrical fittings, gas detectors, and webcams. The value proposition comes from making the user homes smart and reducing operating costs, while ensuring safety and security. The revenue model comes from device sales, subscription sales, installation and service charges, and sale of applications. However, since the device itself can be replicated by competitors in terms of putting together sensors, the company projects its differentiation strategy in terms of service focused. The use of innovative technologies like machine learning based on data gathered over numerous installations helps it get an edge over the competition who may not have that kind of real-life dataset available. Hence, the service centered approach has provided opportunities to innovate at every step of the process, although new products have been introduced at various timelines: thermostats in 2011 to cameras (Nest Cam IQ indoor and Nest Cam IQ outdoor) and security systems (Nest Secure) in 2017. The emphasis on do-it-yourself and off-the shelf offerings along with the fact that multiple devices need to be implemented for a smart home has been a deterrent in adoption since it is cost-prohibitive. The service model is Nest Aware, which offers security monitoring services remotely of its cameras.

As part of industrial automation, GE has launched a Predix platform, which offers a digital, data-driven service and insights to transform industrial operations. It provides actionable insights from asset data and analytics. It also provides an edge to cloud architecture, ensures high control and productivity management, and provides development support to subject matter experts. Their SaaS application ensures asset reliability and availability. The challenges this platform seeks to address is first the untagged sensor data, removing dataset fragmentation, and providing insights and analytics support. Leveraging its domain expertise, GE Predix has been able to offer digital twins (modeling physical with digital), machine learning, distributed intelligence with embedded cyber security, delivering many outputs including scheduling and logistics, connected products, data-based experiences, field service management, analytics, asset performance tracking, and operations efficiency. Hence, the primary business model supported here is offering software as a service and data as a service to a private set of audiences. However, 2017, GE digital supposedly did not bring in the results expected, while in 2013, it was expected to rule the industrial intranet. The reason was that primarily it was driven by internal users such as GE power, GE aviation, and transportation. The value that was expected by customers did not accrue in the real sense because it was just adding technology to existing models and not reinventing a digital centric model and hence the business model was not very successful outside the GE group.

A global flexible packaging company invested into the IoT for operational efficiency, namely in making the inventory control process more automated and instantaneous globally and allowing assets to be interconnected transmitting data for further analytics. In the first phase, OT and IT were integrated by feeding data from sensors to a database system where the data could be used for machine learning to predict failures. In another phase, the inventory items and assets of higher value were tagged with Bluetooth Low Energy (BLE) beacons that could transmit their presence and location information all the time to a digital network, thereby helping in real-time inventory tracking. In the final phase, the IoT sensors were integrated with end machines, which were engineering products to help the customers with proactive service based on data transmitted from the products well in advance before the machines suffered from breakdowns. Also, the product usage information getting transferred provided inputs for future machine design thereby integrating data for new product design.

OLA, India, provides a car sharing service and is an aggregator for cabs. The company is able to monitor vehicle movements and use an app to bring together drivers and cab users. Data is also used to identify and optimize data and help with predictive maintenance. Operational data to some extent can be used by other service providers such as insurers or the government and hence the value can be derived from people even outside the direct interaction circle. The business model operates on a closed and private ecosystem.

Table 4.1 shows the business model building blocks for the four cases.

Google uses the data captured in the digital world to business advantage in advertising by targeted marketing based on analytics of clicks to understand consumer behavior and position solutions accordingly. This is vastly more successful than a traditional mass marketing medium like signboard or newspapers where the target cannot be segregated. The IoT has the potential to further enhance this high-resolution of the customer base and its response since each and every object can now capture response and requests and transfer it without bias to a central server for richer analytics. Also, because of its capability to be uniquely identified, customer and customer whereabouts can be linked for spot-based promotional offers.

TABLE 4.1

Use Case Comparison

Building Blocks	Google Nest	GE Digital	Packaging Company	OLA
Partners	In-house software development, device manufactured in-house	In-house software development, devices from customers	Outsourced, devices from OEM	Outsourced, devices are cars and transport vehicles
Resources	Sensors, Cloud, and Analytics	Sensors, Cloud, and Analytics	Sensors, Cloud, and Analytics	Sensors, Cloud, and Analytics
Activities	Devices developed and expanded through partnerships	Devices of GE, mostly in-house consulting	Devices owned by the company	Fleets equipped with sensors
Value	Efficiency, security, automation	Data-driven analytics	Contextual proximity location-based services	Aggregation of services

Value creation using the IoT can have the following layers as explained with an example:

- *Layer 1* is the physical layer or the object, for example, a Smart TV. Its utility is to provide the primary service for which it is purchased, namely showing channels and programs with sound and other basic features. Its measurement capacity is physically limited to the place it's located, for example, a hall.

- *Layer 2* is the sensor layer, for example, in this case the channels getting played more with the timings of the program being watched with duration. This measures local data and can also be used for providing personalized services such as suggesting similar programs or switching to a channel on a given time when the program gets broadcast without user intervention.

- *Layer 3* refers to the connectivity layer as this data gets transferred to a central server using a Wi-Fi internet connection. Now, this data can become universally accessible and is also the physical object and the sensors can be remotely controlled. In the case of a Smart TV, if the data of a channel being played along with the geocoordinates gets transferred to the cable TV or set top box service provider, it can be used for program enrichment of viewership-based calculations and offers.

- *Layer 4* is analytics where the customer may not see a direct value addition to their primary objective but it gives a wealth of information to the TV manufacturer, the channel content creation team, or to an advertiser who can now identify a target audience interested in a specific content.

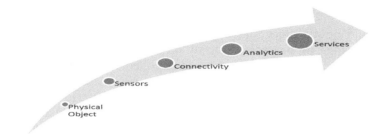

FIGURE 4.2
Five levels of value addition using the IoT in business.

- *Layer 5* is a digital service which can be offered under different business models such as, in this case, it may be some channel subscription offered at a discount or alert service or information service such as a customized offering for subscription to DTH services combining specific channels. It can be an ad-banner integrated with program channels for cross selling, etc. (Figure 4.2).

4.6 The IoT Business Model Based on IT

The value of the IoT can be summarized as the difference between a classical product value and the value derived by the IoT enabling it by the above layers and this value is the price a customer is willing to pay for getting the analytics and this in turn depends on the value on digital services they can get as a business model.

The following is the author's proposition of business models for the IoT (adapting the Gassmann models).

4.6.1 Freemium

This involves installing the IoT devices being sold as implants to existing products, that is, digitally connecting them; some customers may opt for it as a paid service based on value derived later on. An example is Canary (https://canary.is/), which offers home alarm systems at the same price as traditional ones, meeting basic requirement of surveillance while digitally connecting them as well at the same inclusive price. Additional services on a paid basis were offered for a subscription given that the product had the capabilities such as dialing out to a helpline in case of the alarm going off.

4.6.2 Digital Add-On Enhancements

These could be short-term packs that can be purchased to boost existing service offerings, for example, a top-up on a data pack on a mobile phone subscription for 30 minutes. These could be rolled back once the work was completed. For example, the IoT devices can remotely increase the surveillance capacity of the digital equipment for a single day, from 100 footfalls to 1,000 footfalls say in the case of a promotional weekend sale in a mall. Another case would be offering digital insurance for a car being driven by the owner versus being leased out for a commercial purpose on a given day. This will generate premium revenue that is foregone today by excluding commercial usage of a passenger vehicle.

4.6.3 Razor and Blade and Digital Lock-In

This model requires that only original components are compatible with the system and duplicate components cannot be used. The IoT systems can be warned of tampering such as original ink cartridges used in printers. Digital lock-in provides the facility in physical products using a sensor-driven, crypto handshake, this can ensure anytime tracking, keep track of usage and warranty compliances, and prevent counterfeits.

4.6.4 Point of Sales (POS)

IoT devices implanted into vending machines to create enhanced POS machines. A simple wrapper packing of a consumable item integrated with a phone can be used for ordering it through a website.

4.6.5 Direct Selling Business Model or Solution Provider Model using Intelligent Objects Self-Service

This model refers to the ability of things to independently place orders on the Internet. A use case could be a heating system which could order oil refills as soon as a certain level of liquid was noted in the oil tank. This can reduce intermediary costs such as agents or distributors.

4.6.6 Pay Per Use Business Model

With the IoT, it is possible to place a device at the customer end and only bill them for usage. Brother International, a printer company, places printers on lease at the customer's location. Whenever a customer needs to print, they are charged automatically. This model helps customers install more devices knowing full well that they will be charged on a usage basis only.

4.6.7 Digitally Charged Products Business Model

This model is based on adding value to traditional systems by adding technology and thereby selling at a higher cost; for example, smart meters in residences are willingly installed at a higher price because of their prepaid capabilities by users who do not want to go and settle bills on a postpaid traditional model of electricity meters. This has been adopted in many places in India in new age society buildings and planned townships.

4.6.8 Data Sale

Selling sensor-collected data at a fee to service providers is another business model that has evolved. An example would be Streetline, a company that installs sensors on municipal and private properties to detect vacant parking places in order to sell the data collected to interested third parties. This information can be accessed by a car driver free of charge through an app. For city governments, the data processed has a different value by helping plan their physical expenditure to identify parking and then identify offenders; utilization of parking places increases (Schuermans & Vakulenko, 2014) (Table 4.2).

There have been some case studies of failed implementation of the IoT-powered business model such as Bosch, a leading healthcare-product company; Bosch used to run "Health Buddy," a device that monitors the vital signs of patients and records patient inputs. This is targeted at patients with chronic health conditions who can be monitored on a continuous basis from their own homes and data is transmitted to remote medical practitioners. This is an adoption of the "add-on" and "affiliation" business model as this data could be used to service patients and also to pharmaceutical companies to understand effects of their medications. A key reason could be that the customer was not willing to value this add-on service anything more than what the traditional work of the medicine was, that is, to cure them.

There have been some very successful cases, such as Trane that focuses on industrial air conditioning and has changed both its product development practices and the products delivered based on the IoT data and its insights. It adopts the same model and also adds on the crowdsourcing model of data collection to provide insights to new and existing customers on what to buy as per their requirements powered by data gathered by other users of the same device.

Coca-Cola uses data from its freestyle and other vending machines to gain in-sights on their customers' purchasing patterns including where, when, and how they are consuming their products. This also gives it location-specific information leading to selective sales of products and increased vending capacity based on consumption.

TABLE 4.2

The IoT-Derived Business Model

Classical Business Models (Gassmann et al.)	The IoT Business Model Examples
Add-on	Digitally providing add-on like enhanced capacity of smart meter load based on payments from remote
Affiliation	The IoT-measured location-based sales from a smart vending machine
Crowdsourcing	The IoT sensors generating data for common analytics
Customer Loyalty	Measurement of customer loyalty not only on purchase but on continued usage on real-time measurements
Direct Selling	Self-servicing smart objects
Flat rate	Remote usage measurements
Fractionalized Ownership	Also extending to low-value items
Freemium	Providing free digital add-on services powered by the IoT
Push to Pull	Kanban systems, Smart Racks with IoT sensors
Guaranteed Availability	Monitoring the status of production plants or equipment via the internet
Hidden Revenue	Location-specific advertising using the IoT
Leverage Customer Data	Product improvement by manufacturers on the basis of gathered data, for example, cars or televisions
Lock-in	Digital handshake using the IoT sensors and code sharing to prevent competitor products or allow limited competitor products based on consolidated agreements.
Pay per Use	Measurement using the IoT on units of usage
Performance-based Contracting	Measurement using the IoT-based analytics on performance
Razor and Blade	Authentication of part numbers to prevent counterfeits
Self-Service	Objects ordering and replenishing themselves
Solution Provider	Adding more value using smart objects, thereby increasing scalability
Subscription	Time-based subscription and activation, deactivation, for example, DTH
Two-Sided Market	Data consumers and generators integrated for better services such as Fitbit healthcare products

4.7 Challenges and Conclusion

There are numerous challenges entrepreneurs and startups may face while implementing new business models using IoT, some of which are explained below. Previous research is limited on this subject but Wurster (2014) is among the few to categorize the barriers that prevent companies from moving ahead in terms of making money with the IoT. The following are the challenges in this regard.

- Merging of the products and services business may require different level of thinking, for example, a company specialized in products may not have the capability of providing end-user logistics, servicing, or differentiating strategies for production and services since by default while products can be stored, services cannot be stored. One feature of the IoT is the fact that the service portion of the business models outlined is always digital in nature; and there is a risk of services dominating products or altering its characteristics.

- There may be cases where a predominantly physical product merges with the digital product characteristics (either separately or as a combined product offering). In case the cost of the digital portion exceeds the cost of the physical product, decision-making can become very difficult. An error in a product, for example, a car recall due to a faulty airbag can have a devastating effect in the case of the digital portion broadcast errors, which as of now, may get ignored or selectively treated.

- The IoT solutions are very disruptive and have a unique value proposition. Selling them to a customer looking for and used to traditional product lines can be a big challenge in the near future. The kind of premium a customer will pay and its valuation will remain the subject matter of debate.

- Standards are still not uniform and often closed, thereby reducing the number of people who can take advantage of data collected.

- Data ownership is still undecided. Who owns the data for example of a production plant generating data on usage and products being produced? Is it the customer's data since they have purchased the plant or is it the manufacturer's data since they required it to optimize their machine design? Or is it the end consumer's data who needs to know how the product they plan to buy was produced and whether it had manufacturing defects in the first place?

The days of producing a product, gaining a distribution channel, then stocking the shelves waiting for one-time customers to get attracted and then pick up the product is grossly inadequate in the fast-growing IoT market. However, there are still many unanswered issues as the concept is still evolving. The IoT devices adoption has increased, although the prices still remain a barrier, and products initially designed with a strong inclination to the physical world still have a long way to go to be refined to a digital-physical world amalgamation.

References

Alt, R., Zbornik, S. 2002. Integrierte Geschäftsabwicklung mit Electronic Bill Presentment and Payment. In Weinhardt, C., Holtmann, C. (Eds). *E-Commerce: Netze, Märkte, Technologien, Proceedings zur Teilkonferenz der Multikonferenz Wirtschaftsinformatik* (pp. 183–201), Heidelberg: Physica.

Andrejovska, A. 2011. The cost calculation in the manufacturing enterprise. *Transactions of the Universities of Košice*, 3, 7.

Bucherer, E., Eisert, U., Gassmann, O. 2012. Towards systematic business model innovation: Lessons from product innovation management. *Creativity and Innovation Management*, 21, 183–198.

Cusumano, M. A., Gawer, A. 2002. The elements of platform leadership. *MIT Sloan Management Review*, 43(3), 51–58.

Emil, J. 2005. A wireless body area network of intelligent motion sensors for computer assisted physical rehabilitation. *Journal of NeuroEngineering and Rehabilitation*, 2, 6.

Gassmann, O., Enkel, E. 2004. Towards a theory of open innovation: Three core process archetypes. *Proceedings of the R&D Management Conference*, Lisbon, Portugal, July 6–9.

Gershenfeld, N., Krikorian, R., Cohen, D. 2004. The Internet of Things. *Scientific American*, 291, 76–81, doi: 10.1038/scientificamerican1004-76.

Gordijn, J. 2002. Value-based Requirements Engineering, Exploring Innovative e-Commerce Ideas. Dissertation Series No. 2002, 322.

Li, H., Xu, Z.-Z. 2013. Research on business model of Internet of Things based on MOP. *International Asia Conference on Industrial Engineering and Management Innovation*. Springer.

Mazhelis, O. L. 2012. Defining an Internet-of Things ecosystem. In S. Andreev, S. Balandin, Y. Koucheryavy (Eds.), *Internet of Things, Smart Spaces, and Next Generation Networking* (pp. 1–14). Berlin: Springer.

Morris, M. S. 2005. The Entrepreneur's business model: Toward a unified perspective. *Journal of Business Research*, 58(6), 726–735.

Negelmann, B. 2001. *Geschäftsmodell*. München: Vahlens Großes Marketing Lexikon; p. 532.

Osterwalder, A. P. 2004. Clarifying business models: Origins, present, and future of the concept. *Communications of AIS*, 16, 1–25.

Osterwalder, A., Pigneur, Y., Bernarda, G., Smith, A., Papadakos, T. 2015. Value Proposition Design: How to create products and services customers want. *Journal of Business Models*, 3(1), 81–89.

Schuermans, S., Vakulenko, M. 2014. IoT: breaking free from internet and things. Vision Mobile, London. http://www.visionmobile.com/product/iot-breaking-free-internet-things/. Accessed September 20, 2014.

Smith, I. G. 2012. *The Internet of Things*. Halifax: IERC; p. 360, New Horizons.

Sun, Y. et al. 2012. A holistic approach to visualizing business models for the Internet of Things. *Communications in Mobile Computing*, 1, 1–7.

Thestrup, J., Sorensen, T. F., De Bona, M. 2006. Using conceptual modeling and value analysis to identify sustainable business models in industrial services. *ICMB '06. International Conference on Mobile Business*. Copenhagen.

Timmers, P. 1998. Business models for electronic markets. *Journal on Electronic Markets*, 8(2), 3–8.

Westerlund, M. L. 2014. Designing business models for the Internet of Things. *Technology Innovation Management Review*, 4(7), 5–14.

Wu, M. et al. 2010. Research on the architecture of Internet of Things. *Advanced Computer Theory and Engineering(ICACTE) 3rd Intnl Conf On*. IEEE.

Wurster, L. 2014. *Emerging Technology Analysis: Software Licensing and Entitlement Management Is the Key to Monetizing the Internet of Things*. Stamford: Gartner Research Report G00251790.

5

Smart Manufacturing

5.1 Introduction

Industrial automation uses control systems to handle industrial processes and machinery. Smart manufacturing has been defined in multiple ways by agencies such as the Department of Energy and NIST. One of the better definitions (Riddick, 2013) is: "Smart Manufacturing is a data intensive application of information technology at the shop floor level and above to enable intelligent, efficient, and responsive operations." In Germany (Kagermann et al., 2013), the Smart Manufacturing initiative has been referred to as Industry 4.0 or I4.0, while Japan (Nishioka, 2015), USA (SMLC, 2011), India, and Korea (Park, 2015) have their own national programs. Smart manufacturing involves use of a myriad of technologies like big data, predictive analytics, and virtualized process modeling and simulation to create value from data by streamlining data and factory operations (Ransome, 2016). Physical simulation for design and testing may take a lot of time if done in the traditional way using test or pilot plants and going for batch runs (Zhang et al., 2015). However, with digital simulation and modeling, multiple combinations of designs based on material available can be worked out and selected before actually investing in the actual production process. These digital models are used in aerospace research by companies like Boeing and Airbus. Factories stand to benefit with the IoT from operations optimization, predictive monitoring, inventory optimization, and health benefits. With the IoT, many car manufacturers like BMW and Toyota are integrating the full supply chain to identify all supplier parts, going inside the assembly of a car and, thereby, reducing defects and recalls. The IoT can be used to a get a single 360° view of what components are getting deployed where at every point of production. Smart manufacturing can power strategic growth through sustainable means using renewable forms of energy and recycling using a converged approach involving humans, machines, data, products, and digitization. It can help micro monitor to an extent such as in a manufacturing setup like Raytheon, it can be ascertained how many times a screw has been turned, or check and improve quality such as in Merck where vaccine storage conditions are monitored continuously. The IoT can play a major role in safety monitoring by having the IoT-enabled cameras to detect movement of people and machines, or using beacons to track movements and stop or alert people from approaching hazardous locations. Manufacturing products such as white goods—vacuum cleaners, water purifiers, or air conditioners are turning smart, that is, in addition to their physical attribute they are imbibing a digital attribute such as sensors, microprocessors, data storage, and operating systems, which can help with wireless transmission of data including usage and operating parameters. These can be used for further improvement of products such as ABB robotics, and industrial machines can be remotely monitored and adjusted. Further capacities can be controlled remotely on the basis of software only such as Tesla, where everyone has access to the same capacity engine and provisioning of power is done on the basis of the selected model. Smart manufacturing systems will succeed only if we can easily reconfigure factory production and supply networks using systems that can learn from past experience using concepts of artificial intelligence. Also, the entire ecosystem of partners and vendors need to be integrated comprising of all sizes of manufacturers. In existing manufacturing, decisions and diagnostics at a higher level or macrolevel is taken based on undocumented features, while the low-level diagnostics are data based on automated collection and generation. The equipment needed for implementing smart manufacturing systems such as robotic arms or dynamic schedulable machines are complex and require multiple interfaces in a networked environment. Also, integration of a heterogeneous environment, which are proprietary and not open source, presents security risks as multiple protocols are integrated.

5.2 Manufacturing Concerns

The major areas of concern in today's manufacturing organization can be summarized below.

5.2.1 Compliances

With more and more focus on environment, safety, and health it has become necessary to ensure through a system of continuous monitoring, compliances to local regulations. Manufacturers often struggle to provide auditable data points and hence face risk of heavy penalties. Product disposals and checking for alternative input materials require complete visualization of global supply chains since the risk of ensuring suppliers are also compliant falls on the manufacturer. Highly regulated industries like medical devices or chemical plants face regulations such as Universal Device identification and REACH (Registration, evaluation, authorization, and restriction of chemicals). Similarly, food grade materials are subject to various international norms.

5.2.2 Right Product Mix

Manufacturers have to constantly innovate based on customer feedback for their products and services or to bridge the gap created by technology obsolescence. There is also a tradeoff between being systematic and quality on one hand and being the first to market on the other. It is important that each and every product is conforming to specifications and also that their usage patterns are analyzed to bring in improvements, hence companies have to be more systematic about innovation management, and implementing changes on an ongoing basis are necessary for manufacturers.

5.2.3 Skill Gap

Skill alignment to the jobs is a key concern in manufacturing as training shortages and behavioral issues can create bottlenecks, hence more and more automation is being done. However, automation also has challenges of getting more skilled workers for supporting operations.

5.2.4 Shop Floor Productivity

The shop floor has a lot of operations from goods movement, to machine loading, to replacement of parts and logistics shipping. Many unproductive hours can be saved by proper optimization using technology enablers to help with inventory tracking, management and control, preventing idle hours lost due to wrong procedures, and upskilling manpower and upgrading machines at the shop floor.

5.2.5 Energy Monitoring

Energy is a major component of any manufacturing activity in terms of both cost and criticality. There is now a growing emphasis on both conservation and use of sustainable energy for operations. Loss of energy can be a result of carelessness (keep equipment running when not in use), lack of awareness (operating at the right speed and conditions), inefficiency (using machines with duplicate spares or having low energy efficiency machines), or due to other reasons. With a control on energy costs, apart from environmental contribution, companies also benefit economically.

5.2.6 Operational Efficiency of Machines

Keeping machines operational without unproductive breakdowns are essential for the smooth functioning of any plant. Manufacturers often shy away from regular preventive maintenance like replacing cables or go for lower cost products, which can be unsafe and counterproductive in extremely difficult operating conditions. Hence, scheduled plans for upkeep and proper spare parts needs to be in place to ensure machines are operational and there are no revenue losses as a result of unplanned downtime.

5.3 Industry 4.0 and Related Models

The concept of Industry 4.0, which implies automation and data exchange using IT systems and originated from Germany (and has developed similar models across the world with different nomenclature), focuses on the research actions in eight areas, namely standardization and reference architecture aimed at collaboration of partners in a value-driven partnership-focused ecosystem, management of systems with high level of complexity, fast and reliable network infrastructure with broadband connectivity, safety issues in human machine interfaces, autonomous responsibility delegation to employees along with automation, training for the worker so as to handle requirements of the smart factory, regulatory and compliance framework for establishing governance and liabilities for privacy, and resource efficiency in terms of consumption of raw materials and energy. Industry 4.0 also emphasizes on the development of new business strategies based on new business models that are derived as a result of the digital initiatives and may be divergent from the physical product-based selling models.

5.4 Smart Manufacturing

Smart manufacturing incorporates many advanced digital technologies such as robotic arms, automated conveyor belts, contactless transmission and receiving using RFID, Bluetooth, or related technologies, connected supply chain networks with end-to-end visibility of the supply chain from initial raw material provider to the end consumer, and the focus on human-machine collaborative coexistence. The aim is to achieve plant optimization and efficiency and sustainability in the production process and supply chains, which are dynamic and reactive (Chand & Davis, 2010). Smart manufacturing is often used interchangeably with intelligent manufacturing, but researchers have segregated the two in terms of the latter being more technology oriented while the former incorporates nontechnology issues such as business models and human behavior and is more encompassing. The concept of smart manufacturing will interweave the following concepts.

5.4.1 Smart Design

Any manufacturing process will require a recipe or formulation in case of process manufacturing and an assembly design in case of discrete manufacturing as the starting point. Recipes will have ingredient in a fixed proportion, based on defined formulas. A variation in the inputs or recipes has the potential to create new product mixes with different physical and internal properties. The traditional system of creating new product mixes is more by trial and error than data driven or simulation oriented. Similarly, in the case of discrete manufacturing, substituting materials in the standard bill or material in case of nonavailability of spares is usually decided at the spur of the moment without too much simulation and without supporting data of performance. This can lead to higher failure rates or costly future recall of manufactured items. The IoT can help aid this by capturing data at the component level and transmitting information so that behavioral of individual components in a complex real-time working environment can be captured and used for further analysis. Similarly, the IoT-enabled components can be used to simulate the environment, which cannot be physically monitored like the internals of a combustion engine, and then models can be created with substitutes without actual physical changes, reducing the cost of testing out changes in a virtual environment rather than traditional test beds. With technologies like augmented reality and virtual reality, the traditional designing process gets digitized with smart digital prototypes replicating actual physical attributes exactly, leading to run-time modifications to engineering changes and physical realizations.

5.4.2 Smart Machines

Smart machines using intelligence built on data and machine-to-machine communication, such as autonomous cars, self-healing satellites, intelligent robots, etc., are able to solve problems without manual intervention not on the basis of defined rules to work on specific scenarios but dynamically based on training datasets that are evolving as they become subject to more datasets during the course of practical usage. Smart machines such as intelligent personal assistants include Amazon Echo integrated with other IoT-enabled and digitally connected devices to provide machine-controlled environments and experiences (Zhong, 2013). Smart machines should have four capabilities: autonomous operations, secure access storage and transmission of data, digitally connected, and be self-learning with data. CPS-enabled data can transfer collected data to the cloud to synchronize products and services of a machine to an extent that any changes to the physical product can be implemented and then new adapting services provisioned remotely. CPS-based smart machine tools can be used to produce physical products since they integrate smart machines, warehouses, and production facilities end-to-end digitally.

5.4.3 Monitoring

Monitoring of operations and designing the controls and ensuring scheduling operations and maintenance activities through the use of embedded sensors is a significant advantage of using the IoT (Janak & Hadas, 2015). Inventory control using Bluetooth Low Energy (BLE) beacons by enforcing real-time and continuous tracking is an important use case as explained in the case of a packaging industry where roll movements are tracked at real time using beacons that transmit not only real-time location information but also give out contextual information (Ramakrishnan & Gaur, 2016a). Smart inventory monitoring and management can help in customer service and cost optimization.

5.4.4 Control

Smart controls help in establishing high resolution production controls over cyber physical systems, and have to be extended to the cloud to ensure they can be exercised remotely from anywhere, thereby enabling viewing and control from any location and by any concerned stakeholder in the value chain, bringing in collaborative controls. These work with robotic assembly lines and smart machines (Stich, 2015).

5.4.5 Dynamic and On-Demand Scheduling

Using data captured from the IoT sensors and machines, it would be possible to use extrapolation and algorithms for advanced-level decision-making. Real-time scheduling can be useful in a virtual factories model (Sanjay, 2001), which is an integrated simulation model with advanced decision-support capabilities where real-time order fulfillment and execution is of prime importance with the use of distributed models (Marzband et al., 2016). Data inputs can be used for interparty resolutions involved using different methods for group decision-making (Güleryüz, 2016).

5.4.6 Collaboration

The IoT and smart manufacturing extends traditional systems and information transparently across different participants in the value chain as competitive advantage is derived more from plant capacities and marketing channel strength to data, value addition, and shared facilities and services. As virtual facilities become more readily available and new innovation and new product development is driving business growth, collaboration is becoming the new norm as compared to the traditional practice of keeping operational data in strict secrecy. Collaboration helps each participant get a full view across the value chain, for example a leading automobile company with sensors implemented in wheels not only gets data on the utilization of the wheels and its effect on engines for its self-consumption, but also shares this data with its tire manufacturers and wheel rim manufacturers to get a synergistic improvement focused on all concerned, through a convergence of the IoT and unified communication systems. This has also helped combine things that are nature wise connected such as cars to things which are disparate and not connected such as wheels. Also, the assessment can be done in real time since data is continuously transmitted and also a fast response time will be needed for this volume of data. This helps one of the biggest challenges of process manufacturers, which is to bring process flexibility and adapt to changes in demands.

5.4.7 Digitized Products

Products traditionally had a physical attribute only and hence operated independently. The downside was that information on consumption, usage, or performance was not available to the manufacturer and these products had to be serviced or analyzed by an onsite visit by the technician, which was a costly affair. Also, limited data was stored locally since the digital aspect was negligent. While earlier barcoded products and then RFID stickers introduced some kind of passive communication, current digitization helps remotely control and reconfigure products to provide better services or introduce new innovations that can transform manufacturing (Rockstorm, 2018). The IoT-enabled products bring in new business models (Paul, 2015).

5.4.8 The IoT-Driven Process

Future internet will see humans being replaced by objects as their primary users in terms of volume of data being transferred and the time spent in transmitting data. Traditional digital systems such as ERP can benefit from nonhuman actors such as machines, objects, and robots taking over the role of process resources and interacting with the physical environment in ways identical to what humans are doing today for individual tasks. These can also lead to modification of existing business process and it is necessary that the IoT devices get identified as a process resource in a traditional digital such as ERP and new business process life cycle models (Scheer, 1992).

5.4.9 The IoT and People

People safety and security is one of the biggest concern areas in manufacturing, especially in conditions of high temperature (boilers and furnace), radioactivity and magnetic rays (centrifuge and power generation), high pressure (compressors), fast moving assembly lines (chains and shafts), or high decibel sounds (vibrators and motors) that are highly accident prone. The IoT-enabled machines and sensors can help detect human proximity to such areas, CCTV with infrared detection and motion detection and warning alarms can also warn about the presence of humans in time. Office buildings can become more people friendly and environmentally friendly by detection and adjusting to preferences of people through prestored memories. Right from facial detection to voice synthesis, personal assistants can be of immense help in business leading to security systems in business environments. A demo project conducted by IBM in Munich helped factory workers to become location aware and also transmit their location information. This was helpful when machines needed repairing by using an alarm to alert workers before there was overheating and hot metal spill.

5.4.10 Scalable IT Systems

The IoT enablement in manufacturing enterprises requires substantial IT infrastructure for integrating Operational Technology (OT) with Digital Technology (IT). Substantial compute power, memory, storage for data, cloud computing, and next-generation computer networks are needed for any manufacturing organization to implement the IoT. A major challenge is in integrating OT, which may not be compatible with TCP/IP protocols and may be proprietary to connect to TCP networks. Manufacturing enterprise systems have numerous design and operational requirements, and numerous design variables, hence IT systems for data capturing has to scale else performance of MES systems will become very slow. MES requires decision-making inputs from IT systems and similarly data from machines flows through MES to IT systems for analytics while maintaining data quality. Manufacturing systems have evolved over time from craft systems, British systems to American systems. Post that, new concepts of lean production, FMS, and sustainable development has taken over and at each stage the number of variables have increased multifold (Bi et al., 2011). There have been literatures on conceptual model and enterprise models (Franke et al., 2013); however, most of them are static and cannot be changed to adopt to new requirements such as holonic manufacturing or agent-based manufacturing. Agile systems were developed with an objective of integrating with enterprise IT systems, to increase system adaption. Any IT system to integrate with manufacturing systems need to be able to address the complexity, dynamics, and uncertainties and be able to integrate virtual environments. Merging IT systems with MES also faces challenges on account of ubiquitous compute requirements, vertical integration between all departments right from design to production to sales and logistics, horizontal integration between suppliers, manufacturers, and customers, and the need to work with multiple sources of proprietary data (Bi et al., 2008).

5.4.11 Financial Feasibility

The IoT applications in manufacturing can provide easy-to-recognize high-RoI applications like preventive maintenance or emission controls and energy conservation that can be measured under key indicators such as reduce costly downtimes, improve productivity and reduce unit cost of production, help simulate newer business models, improve and convert marketing opportunity, regulatory compliances, and improve operating margins. Also, the cost of solution investment in terms of the capital expenditure and payback period is necessary; however, with lower cost OPEX models using cloud-based infrastructure, such as software as a service (SaaS) models, which are use and pay, it becomes easy to get solutions implemented at low cost. Also, post sales service costs can be reduced with location-based services and new product developments catering to the actual needs assessed objectively by data captured using IoT systems. Financial evaluation is compounded by the number of peculiarities in the IoT such as the number of systems and end points affected and part of the solution, multiple vendors, and entities including network providers, solutions providers, hardware providers, financial drivers for adoption, and hence various measures such as project profitability, cost benefits, return on investment and payback period apart from intangibles like user adoption rate, technology competitiveness, product digitization, and its benefits.

5.4.12 Autonomous Functioning Driven by Machine Data

Machine data can help power artificial intelligence in manufacturing systems that can self-heal, self-repair, and self-program for more productivity and efficiency as the objectives. Robotics can learn from data and hence traditional semantics of if-else-then logics are no longer needed as systems learn from data getting captured, which are used as training datasets. These get evolved with time as more and more data get captured during the course of normal operations. One case is the iRobot vacuum cleaners that can help map home layouts and clean the room with efficient movement pattern or even dock autonomously when it gets discharged. Deep learning takes this to a next level with complex structures and multiple processing layers to model abstractions from data. Most of the IoT devices have built-in ambient intelligence, which refers to electronic movements as a response to people's presence. Machine data can help with establishing management controls such as when inventory is to be replenished, what is the minimum stock order, how many hours machines can work before a specific part gets a breakdown, what is the optimum speed at which machines can be operated, what is the pressure a motor can be subject to, and similar measurements.

5.5 Smart Manufacturing: Indian Case Study

This section highlights how an Indian company with diversified business interests in chemicals, packaging, engineering, and other divisions globally adopted the IoT across various stages of its business and is involved in both process and discrete manufacturing operations. In the process operations, raw material silos dispense the exact quality of raw materials in solid form in weight while the liquid raw materials are controlled by the IoT-enabled flow meters in volume depending on the density. The process is initiated by marketing teams that traditionally used to feed data using connected mobile phones but now follow a collaborative ecosystem where the orders are automatically picked up from warehouses that sense data in customer premises using the IoT-enabled racks and bins. Once the data comes to the digital enterprise resource planning system and is confirmed by a human interface, the ERP communicates with the manufacturing execution system (MES), which in turn communicates to the PLC of the machine (jumbo machine) for initiating a batch based on the flow meters. The silos are then activated to release the required quantity of raw material for the batch (Ramakrishnan & Gaur, 2016b). The IoT sensors through an integrated SCADA controller in the jumbo machine not only transmit data about the quantity of output being generated but also check the quality using laser lights where the attributes like opacity, thickness, and coating levels are measured. The output data serves as input data for the secondary batch that consists of small line slitters, which are, again, equipped with the IoT sensors to further capture output parameters. The identification between the machines is done using BLE stickers attached to the products right from the time of creation and at every stage of creation, which is the unique identity. These stickers are scanned and detected from a controllable distance of as less as 2 feet to as much as 100 meters. So, the near proximity readers attached to secondary slitters detect the material only at a distance of a maximum of 2 feet, which is when the items have been loaded mechanically into the slitting machine's arms. However using the distance mode, inventory lying on the shop floor or warehouse or works area can be detected anytime by remote monitoring Bluetooth-enabled scanning devices or handheld devices leading to easy detection and instant cycle counting (Ramakrishnan & Gaur, 2016a). Finally, the IoT-integrated robotic arm gets instructions on the weight of the items to be packed along with the sizes measured by a digital ruler equipped with a laser. The robotic arm picks up the items and places them on either of the conveyor belts based on whether it is meant to be dispatched to the end customer for final shipping "make to order" or if it has to be kept as "make to stock," which has to be sent to the warehouse. Finally, handheld scanners are used to load the items to freight truck carriers and final dispatch happens from the factory premises. Before digitization, the operations were carried out manually right from order booking to feeding data into the SCADA for plant operation to slitting and final dispatch. The system was prone to errors by human interfaces and time taken to complete operations was high, leading to slower customer service and costly recalls.

In yet another scenario, one of the divisions is in the process of manufacturing engineered machines. The machines are not only assembled and sold but also serviced on an annual contract basis. Many of the customers can afford just one machine and their entire production chain depends on the smooth functioning of the machines. Many of the failures depends on one of the crucial mechanical components from over 1,000 items that are assembled to create a machine. Before digitization of the assets, internal operations were ineffective on two terms: one, to identify and assemble materials or their substitutes, and second, to ensure that parts that had different replacement guarantees could be uniquely identified against each order. With the introduction of the IoT, the machine and its components could support proactive identification of breakdowns by transmitting data using Wi-Fi or other connectivity when values used to cross certain thresholds. Comprehensive logging and remote monitoring of material enabled by digital channels also helped repair with shorter and more planned down times and helped identify common failure points. This, along with usage data, helped the manufacturer get insights for new product development. The notable benefits being a reduction on down time leading to better customer service, better inventory control of the numerous discrete components, new product innovation driven by data, and lower cost of operations due to optimization. The IoT-enabled products can also be provisioned or deprovisioned on demand and can help introduce new business models totally different from the earlier physical product sales, thereby introducing diversification and competitive advantage based on digital models (Figure 5.1).

The IoT devices in all of these scenarios exhibited some key traits such as integrated networks with wireless sensor networks, thereby allowing seamless communication with enterprise systems allowing for storing and analyzing captured data, dynamic and scalable architecture built on an integration bridge, which could be reconfigured as machines get reprogrammed or updated, support cloud-based architectures, and also work in collaboration with the human elements.

The company has worked on a conceptual collaborative creation model for new product development where end customers are themselves contributing to the development of new products by providing usage inputs and preference inputs. This can be explained in four layers: (1) physical product, which in this

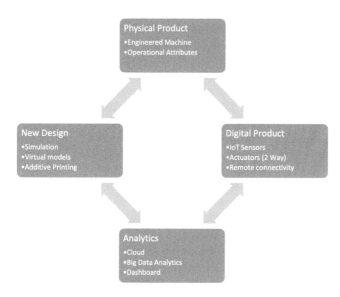

FIGURE 5.1
New product development conceptual model—the IoT.

case is the engineered machine, (2) digital product, which includes the IoT sensor responsible for measurements including actuators and motors with digital capabilities, (3) analytics, which is on the cloud, and (4) new product design, which is based on virtual simulation models and then translating them to physical models using additive manufacturing.

Additive manufacturing helps translate designs to physical models using multiple-layer additive printing using different types of substrates. It is possible to design a specific gear or level using additive manufacturing and substitute it in physical models before mass production to see whether it has any substantial impact on machine efficiency and also the cost of machine assembly. Additive manufacturing is useful in rapid prototyping using either plastic, metals, or composites and is now being increasingly used in serial production. It helps organizations create distinctive profiles, keeping in mind new customer benefits, costs, and sustainability objectives.

5.6 Conclusion

Traditionally, machines are designed only keeping the unit production efficiency, and planned production logics objective. These are not tailor-made for digitization, presenting working challenges in the adoption of smart manufacturing enabled by the IoT sensors at the shop floor. Data capture becomes a by-product as people are more concerned about actual production, thereby data is lost and, with it, information that can be used for product performance monitoring or machine performance control also gets lost. With smart manufacturing and automation, data gets continuously captured, retrieved, and transmitted giving insights to gauge performance or predict failures. Smart sensors can capture physical attributes like temperature, vibrations, humidity, environmental conditions, gas and corrosive elements, speed, and impact, which can capture information at end points and help create virtual models. Manufacturing companies have a challenge of visualizing new services or innovating new product or recipe mixes or testing a new bill of materials with substitutes, without incurring charges on account of failed trials and investing in pilot plants. Smart manufacturing using the digital IoT can help model and simulate in a virtual environment a myriad of product combinations and also test out new services, while allowing these changes in design and structure to easily propagate to the end users by remote configuration and provisioning. Decision-making will become much easier due to multiple data points being captured and analytics supporting and mined from these data points. Future research should be based on two areas that are currently not adequately supported by literature: firstly, augmented-reality-enabled real-time visibility in manufacturing machines by using simulation so that machines working can be visualized in different environments, and secondly, cyber-virtualization modeling in a cloud environment leading to provisioning of intelligently created services that can be used in a collaborative manufacturing ecosystem consisting of all players from manufacturers to aggregators. Smart manufacturing can bring in customizability to create lot sizes of one, increased productivity, data-driven analytics, reduce lead time to market, and better compliance, worker, and environmental safety. This chapter provides insights using a case study where the IoT is used right from design to post sales servicing and also suggests a conceptual framework for smart manufacturing while also presenting challenges and limitations in the current approach and future research directions.

References

Bi, Z. M., Lang, S. Verner, M. Orban, P. 2008. Development of reconfigurable machines. *The International Journal of Advanced Manufacturing Technology*, 39, 1227–1251.

Bi, Z. M., Lang, S. Verner, M. Orban, P. 2011. Revisit system architecture for sustainable manufacturing. *J-SustaiN*, 3(9), 1323–1340.

Chand, S., Davis, J. F. 2010. What is smart manufacturing. *Time Magazine Wrapper*, 28–33.

Franke, J., Charoy, F., El Khoury, P. 2013. Framework for coordination of activities in dynamic situation. *Enterprise IS*, 7(1), 33–60.

Güleryüz, G. B. 2016. Multi criteria group decision making approach for smart phone selection using intutionistic fuzzy TOPIS. *International Journal of Computational*, 9, 709–725.

Janak, L., Hadas, Z. 2015. Machine tool health and usage monitoring system: An initial analyses. *MM Science Journal*, 794–798, doi: 10.17973/MMSJ.2015_12_201564.

Kagermann, H., Wahlster, W., Helbig, J. 2013. *Umsetzungsempfehlungen für das Zukunftsprojekt Industrie 4.0: Deutschlands Zukunft als Produktionsstandort sichern.* Frankfurt: Geschäftsstelle der Plattform Industrie 4.0.

Marzband, M., Parhizi, N., Savaghebi, M., Guerrero, J. M. 2016. Distributed smart decision-making for a multimicrogrid system based on a hierarchical interactive architecture. *IEEE Transactions on Energy Conversion*, 31, 637–648.

Nishioka, Y. 2015. *Industrial Value Chain Initiative for Smart Manufacturing.* Tokyo, Japan.

Park, J. 2015. *Korea Smart Factory Program.* Tokyo, Japan.

Paul, J. A. 2015. Bottom-up approach based on Internet of Things for order fulfillment in a collaborative warehousing environment. *International Journal of Production Economics*, 159, 29–40.

Ramakrishnan, R., Gaur, L. 2016a. Feasibility and efficacy of BLE beacon IoT devices in inventory management at the shop floor. *International Journal of Electrical and Computer Engineering*, 6(5), 2362–2368.

Ramakrishnan, R., Gaur, L. 2016b. Application of Internet of Things (IoT) for smart process manufacturing in Indian packaging industry. *Information Systems Design and Intelligent Applications*, 435, 339–346.

Ransome, J. W. 2016. *Cloud Computing: Implementation, Management, and Security.* CRC Press, Boca Raton.

Riddick, E. W. 2013. *Panel on Enabling Smart Manufacturing.* USA: State College.

Rockstorm, J. 2018. *The Fourth Industrial Revolution: What Is Means, How to Respond.* World Economic Forum.

Sanjay, J. N. 2001. Virtual factory: An integrated approach to manufacturing systems modeling. *International Journal of Operations & Production Management*, 21, 594–608.

Scheer, A.-W. 1992. *Architecture of Integrated Information Systems: Foundations of Enterprise Modelling.* Berlin–Heidelberg: Springer-Verlag.

SMLC. 2011. *Smart Manufacturing Leadership Coalition, Implementing 21st Century Smart Manufacturing.* Washington DC: Workshop Summary Report.

Stich, V. N. 2015. Cyber physical production control: Transparency and high resolution in production control. *IFIP Advances in Information and Communication Technology*, 459, 308–315.

Zhang, Y. F., Zhang, G., Wang, J.-Q., Sun, S. 2015. Real-time information capturing and integration framework of the internet of manufacturing things. *International Journal of Computer Integrated Manufacturing*, 28, 811–822.

Zhong, R. Y. 2013. RFID-enabled real-time manufacturing execution system for mass-customization production. *Robotics and Computer-Integrated Manufacturing*, 29, 283–292.

6

Energy Consumption

6.1 Introduction

Energy is the most important component for any manufacturing process (Chand & Davis, 2010) to either convert materials from a raw to finished state or provide support with heating cooling functions. Manufacturing results in almost 84% of energy-related CO_2 production and 90% of industrial energy consumption (Scipper, 2009).

A sustainable energy program has become a necessity due to increasing pressure on reducing fossil-based fuel and preventing global warming. The growth rate of industries is increasing the demand for affordable and clear energy. Energy-efficient manufacturing or green manufacturing implies that goods be manufactured with minimal energy consumption, which can be facilitated by having higher energy efficiency ratio machines and preventing energy wastage due to faulty components or human negligence and the IoT can play an important role (Haller, 2009).

Smart metering solutions adopted in residential complexes have faced challenges of data transmission across networks, tamper proofing of readings, no two-way communication, and ensuring accuracy of readings with volatile loads. Smart sensors and smart meters can help increase visibility and awareness on energy consumption, saving costs in the face of spiraling energy prices (Bunse et al., 2011). Green products and manufacturing emphasize on creating products that consume less energy when used by customers (Garetti, 2012). Organizations are moving toward an innovative process in manufacturing as well as energy monitoring and management (Weinert et al., 2011).

Literature exists on the steps taken by discrete manufacturers to reduce pollution and ensure energy efficient production (Duflou, 2012). Energy management programs involve a blend of strategy, technology, and management decisions, and are multidisciplinary and cross-functional. However, in the case of manufacturing, these face numerous challenges because of a variety of energy practices across multiple processes each having its own set or attributes. One innovative case of autonomous energy management in residential areas has been explained (Ramakrishnan & Gaur,

2016a,b,c) where all connected objects such as the consumer of electricity (fans, air conditioners, hospital equipment's, lighting, etc.) are IoT-enabled, all producers (electric generation stations, grids) are IoT-enabled, and all distributors are IoT-enabled (transformers, substations, distribution stations). These consumer objects will work in a hierarchical chain of command autonomously requesting for load sanction from its immediate supervisor (smart meters or transformers), which in turn will request the producers all in real time. This kind of arrangement will help monitor traffic, prioritize requests, and prevent a blackout due to overload. This can also bring a dynamic and differential metering slab-based system where high priority items such as medical equipment can be given subsidized power while low priority items such as lightings of an entertainment mall can be given power at higher tariff rates. The same goes for power cuts when there is a shortage of supply as compared to peak demand. The transformer can instruct the smart meter to switch off nonessential power consumers like air conditioners or instruct a specific set of consumer category like malls, while allowing others to continue. Traditionally, such microlevel management of energy is not possible, where in load shedding it is done location-wise or timewise, which may not be a very rational way.

Similar energy management structures have been created for factories based on consumption data and employing methods like load balancing or proactive maintenance (Kannan, 2003). A number of models are presented in existing literature for energy end users monitoring and optimization, including for the metal industry (Gordic, 2010). As defined in ISO 50001, for improving energy efficiency it is necessary to monitor and analyze energy consumption. The inability to set up submeters or measuring equipment is also a barrier to energy efficiency in manufacturing companies (Rohdin et al., 2006). Hence, manufacturing machines at the shop floor is not metered in a continuous basis (Müller, 2009) leading to a very poor level of energy controls in manufacturing operations (Garetti, 2012).

Currently, energy consumption rates are calculated, which can be based on historical data or on the basis of production factors like output quantity. However, these are fraught with issues as they are subject to accuracy issues, bias introduced due to fluctuating patterns; hence, for real-time monitoring, sensors have to be integrated (Bunse et al., 2011), and power monitoring has to be part of different levels of operations at the shop floor, to collect data on consumption.

Metering can be done at three levels: at the factory level using meters installed at the point of main connection, at individual line levels for calculating production costs in relation to energy for the entire line constituting a set of diverse operations, and at the machine level where each individual machine can be monitored for energy consumed. The metrics that are measured include idle-time consumption, batch-wise consumption during the machine-run cycle, and utility measures measuring total energy input by the service provider. With the IoT-enabled meters, manual meter reading will no longer be required as smart-connected meters can transfer information directly to a metering panel. Hence, the benefits of the smart IoT meters can be listed as continuous monitoring of multiple points, automated meter reading, and transmission on a real-time basis, capable of measuring multiple parameters such as voltage, current, load, temperature, and time of use and idle time.

6.2 Energy Management Practices

As consumers are becoming more concerned on whether their products come through energy efficient means of production, along with the supply chain operations such as logistics and up to recycling, it has become imperative to adopt energy management practices to protect the brand (Bunse et al., 2011). Any consumer or industrial product goes through three broad phases: manufacturing, followed by service once it is commissioned or sold, followed by recycling once its utility is expired. Hence, a total energy management practice needs to integrate all three cycles; whereas, the current focus is primarily on the manufacturing stage, which is also the largest of the three.

However, as of today, there is an energy efficiency gap more so in developing countries and companies are not fully aware of the potential economic benefits of adopting good practices (Groot, 2001). Numerous energy management solutions for the industrial sector exists such as Wi-Lem, Watts Up, Epi Sensor, GE, and Mitsubishi. There are also some energy measurement software providers such as Google, EFT-energy, and Resource Kraft.

Energy observations seek to find answers to different business questions such as identifying process schedules, which reduce total time and also take minimum energy for completing a task. It also helps identify indirect energy that can then be apportioned to process and machines. Further trading of energy can happen by identifying internal opportunities, after collating data of surplus or deficit, in a given fixed time period like a day or month and then selecting buyers as per best preferences. Trading scenarios also seek to address and identify if energy can be reallocated to the supply chain for better optimization and reduced costs, and create stable models to redistribute energy among partners. There have been studies conducted in ensuring energy management in machine workshops (Chen, 2018). However, most of the existing literature concentrate on machines when it comes to energy monitoring (Vijayaraghavan & Dornfeld, 2010), and energy saving can be achieved by a streamline production process making it more energy efficient or by using novel methods in monitoring.

Earlier research has addressed challenges in energy efficiency for production machines and factory buildings (Menz, 2015). The Bosch Rexroth

Efficiency Workbench (EWB) was developed to analyze and co-relate process data of a machine against its physical production process based on which cycle time optimizations are possible. The initial start of machines draws the highest energy due to simultaneous spindle acceleration from a stopped state and for peripheral systems like a boiler or chiller to reach the desired temperature state. If it could be possible to schedule machines to start one after another while ensuring that before start of process all available resources are ready and mobilized, it may help downsize the initial power generation capacity as well. A very simple example would be a set of window air conditioners of 1.5-ton capacity having a starting load of 4 KW while its running cost is 1.5 KW with the compressor on and in other cases less than 600 W; however, a firm has to size for all the air conditioners considering 4 KW as the peak load demand. However, if the equipment could be so programmed to start simultaneously one after the other, and the compressors could work synchronously in harmony controlled by the connected IoT systems, the energy requirement could be reduced to as low as 600 W plus a spare of 1 KW for the compressor running time. This would be possible only if the temperature monitoring is done on a real-time basis and integrated with an energy monitoring and control system that can then issue a stop/start command to individual equipment.

Another aspect is the use of energy management for the monitoring of works quality. Most of the process manufacturing is a set of small steps of production tasks (Franke et al., 2013) and a flaw in any of these intermediate steps may result in economic loss, and usually post fault diagnosis results in rework and has limited utility. Monitoring systems using the IoT can be used to capture direct data such as speed of the motor or derived data such as energy consumed, which is correlated to the motor speed. In some areas such as high temperature boilers, etc., where sensors cannot be directly used, it may be necessary to monitor only derived data. There has been a good example of using energy measurement for deriving flaws in a work piece of a turning machine. A tool cutting through the spindle experiences a higher energy pull when the bore reaches the spindle as compared to plain air.

6.3 The IoT for Energy Management

The following section deals with general energy monitoring system architecture, which derives support from earlier energy-aware systems based on IT, which could provide personalized energy to building occupants (Fotopoulou et al., 2017). It consists of multiple layers including sensors, measuring meters, network connectivity, gateways to support TCPIP-non-TCP conversions, and cloud-based analytics. The value addition by the IoT-enabling energy-capturing solutions ensures that energy consumption and demand can be monitored at different levels, both as segregated and aggregated.

Any conceptual model integrating the IoT and energy management has to integrate the following layers: shop floor monitoring of energy, estimations of energy, trading energy blocks, and data capture blocks. Here, we propose an IT-driven energy management model that can integrate diverse data sources—MES, APS, and ERP (Watson et al., 2010). This will essentially be a distributed service-oriented architecture and independent of industrial application.

The main elements of this model include "smart physical objects" that can sense information and transmit data using embedded sensors relating to energy parameters, and refer to plant machinery, conveyor belts, slitting machines, or hydraulic-powered machines used in compression or expansion or bending, or boilers, chillers, and air conditioners. The "IoT enabling components" are added to these machines to impart them digital attributes other than their original physical attributes, thereby making them connected objects and allowing two-way transmission of data and remote monitoring as well as control. "Manual methods" of energy capture those which are not the IoT-enabled machines and refer to analog meters or PLC/SCADA devices attached to machines where operators have to take readings and enter in a system. "End product" materials information store energy footprint and carbon footprint data, thereby enabling the carbon rating and ensuring they are energy efficiency rated. "Process energy estimators" refer to a set of tasks and machines forming part of the end-to-end process, energy consumption estimator, which serves as the middleware to aggregate and analyzes data with the ability to co-relate to actual production data. "Production with energy focus planner" is a use case actor who does the planning on how to fulfill orders and includes the energy component focus with traditional material and production planning. "Production flow optimizers" are algorithms using historical data for prediction and incorporating a software-based expert system or artificial intelligence system to take decisions on what to produce, what form of energy to use, and what process to adopt with maximum productivity and energy efficiency as the driving constraints. "Energy trader" is an ecosystem for buyers to buy low cost and environmentally friendly energy as per local regulations and also sells extra energy available and trades in carbon credits (Figure 6.1).

From the IS viewpoint, there will be multiple layers supporting this kind of organization starting from the data acquisition layer that has to be interoperable with heterogeneous data sources, then a fog computing layer at the meter level, transportation layer to transmit relevant calculated data, a cloud layer for storing the huge volumes of data, an analytical layer that does further analysis on the data to help decision-making, and a reporting layer that can be used by energy traders to view credits and work on an energy exchange system ensuring full transparency (Figure 6.2).

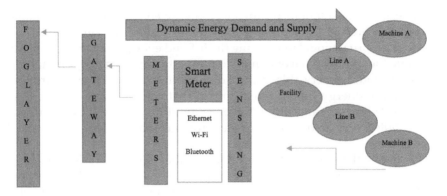

FIGURE 6.1
On premise section of energy monitoring system framework.

FIGURE 6.2
Six-layer IS model for the IoT-driven energy management systems.

6.4 Industrial Case Study

Using case study methods, authors have investigated the energy requirements of a manufacturing company in India, which has adopted the IoT-based shop floor metering system up to the machine level for its round-the-clock energy-intensive manufacturing processes. Before the start of this project, the main objectives of the energy management program were defined as follows:

- Energy costs were around 10% of the cost of production and different sources of energy generation existed such as generators based on natural gas, generators based on diesel and furnace oil, and state-owned power with high-capacity dedicated lines. The prices and advantages of each of these were different and the problem was compounded because all of them had to be used to some extent (due to contractual commitments) and prices of hydrocarbon products used to vary almost daily as uncontrollable factors.

- Since each machine was subject to different types of process and subprocess, such as chemical coating, metallization, thermal treatment, hence, even at machine level for different batches, the energy requirements could be varying and had to be accurately measured.

- Hence, a system was needed to have an extremely granular layer of energy consumption capture at the machine level, at the process level, and at the department and facilities level and since many of these processes were supported by ancillary machines, which also were power consumers such as chillers and boilers, it was necessary that all the data was collated in real time to arrive at power distribution.

- Another objective was to develop and implement a tailor-made energy efficiency program that could meet the customized business need requirements and that could start right from the production planning stage and operations scheduling. It was also desired if the system could use IT for generating complex computational logics, which could be used with constraints like fast production, lowest energy consumption, and planning based on different regulatory and demographic footprints of different states.

- Yet another requirement was to have energy consumption parameters identified with substantiated data per finished product level so that it could be mentioned for end consumers of the product to see and validate the energy spent on production.

- The program was required to be vertically and horizontally integrated across multiple departments within the firm and also with vendors whose data could either be integrated directly from their systems or be manually fed.

- Another driving factor was for reducing the carbon footprint by not only monitoring energy consumption, and controlling it via optimization, but also through forecasting and trading. Hence, systems were needed that could enable more accurate forecasting and provide valid data for carbon trading (Figure 6.3).

The system needed to implement the above objectives and had to get information from multiple sources such as MES or even SCADA and hence, the IoT sensors and the IoT gateways were needed to ensure the machines were digitally equipped to capture and transmit information to a common database. Product information needed to be picked up from a product life cycle management system (PLM), while inputs of quantity and stock used to come from MRP modules, part of an ERP system. The study compared the rated power and the actual power consumed for each of these physical processes to find out if the individual machines and the total machines were performing as expected for each duration of run. There were some dummy runs made to check the energy efficiency during idling speed.

The following five subsystems were identified as sources of information to be fed into the energy management system:

1. Data relating to operational process including orders, products, materials.
2. Data relating to production such as machine working hours, process and task details, and tracking product movements.

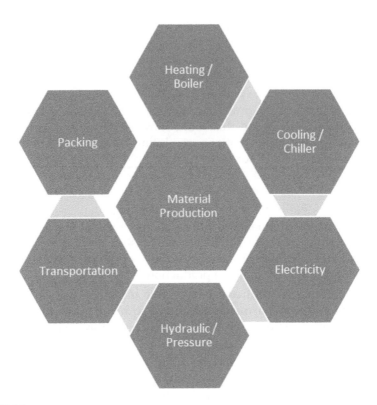

FIGURE 6.3
Energy consumption processes.

3. Energy monitoring using sensors at individual machine levels, integrated with transmission gateways and smart meters and energy vectors, which are a set of average values of energy consumption.

4. Carbon profiling and environmental data capture for each product and process.

5. Requests for carbon credits by partners or energy exchange programs.

This set of user requirements were met with an integrated-systems approach wherein firstly all machines were integrated to have all components of energy monitoring enabled along with measures that could be used for corroboration such as operating speed of motors, voltage, and current spikes and vibrations along with operating temperature. The thumb rule was to ensure that the sum total of energy consumed matches the energy produced and there was minimum unbilled energy flowing in the system. Unbilled energy is the energy that has been generated but has not been consumed or stored and hence is lost. The unbilled energy before the start of the program was as high as 40%, in spite of best human planning, since the energy requirements were measured one time and not on a continuous basis and there was no real-time data available. The energy monitoring was done at three levels: firstly at the process level, that is, the indivisible process, for example, the heating of a boiler to create melt from plastic granules was tightly integrated as a process; the second level covers inter-process and intra-department monitoring such as between utilities and manufacturing; while the third level refers to an aggregated level, where values are at an estimated level, such as future production order. Energy monitoring at the process level can be used to relate business data with energy data and understand if process changes are required to implement better energy management functions. Further comparison of energy across the process can be used to identify differences between entities. The typical data model needs to capture information relating to the ordering process, departments, BOM, and materials and machines (Eleni, 2014). The monitoring can be done by energy metrics or analytics, helping to address decision problems such as scheduling in multiple steps, production time, and energy costs per process step and per production step.

The study resulted in the adoption of energy efficiency measures such as converting units to use energy based on actual usage rather than fixed usage, converting motors and lights to higher EER ratings by replacing them, ensuring optimal temperature in the plant that is neither too cold or too hot, since both would reduce efficiency while increasing energy costs, designing warehouses with small sizes to reduce cost of cooling, enabling sleep mode in a variety of equipment such as a hydraulic pump, conveyor belt, and pneumatic pump, and adoption of motion-based power activators, avoiding the waiting time between machine material exchange, thereby reducing idle time using scientific and data-driven methods.

With the kind of data available, it is also proposed in the future to adopt deep-learning neurons-based system-simulated information, which can identify associations between energy and other variables.

6.5 Conclusion

ICT solutions integrated with the IoT sensors can go a long way in ensuring energy saving, by identifying up to machine levels and even submachine component levels the energy requirements needed. This chapter discusses the considerations a decision-maker of industrial manufacturing – discrete or process—will need to keep in view to reduce the energy footprint and, hence, derive better economies. The IoT solutions are flexible, cost-effective, and can help monitor and track energy changes at the lowest level of consumption and, hence, should be an integral part of any energy conservation and measurement program. It also identifies the benefits that can be derived from such a solution, including a process of continuous improvement. A general framework is also defined that can help implement progressive and advanced energy management practices. With rising prices of energy and fuel, manufacturing industries, specifically, have a big energy cost of their overall manufacturing costs and, therefore, have a mandate to reducing their energy footprint. Analytics on collected data can help in insights on peak and low consumption of power, helping decision-makers replace obsolete equipment and go for new energy-efficient models or look at a mix of renewable energy practices as part of their sustainability programs as a catalyst for innovation and to ensure environmental concerns and adhere to regulations and legislations.

References

Bunse, K. V., Vodicka, M., Schönsleben, P., Brülhart, M. 2011. Integrating energy efficiency performance in production management: Gap analysis between industrial needs and scientific literature. *Journal of Cleaner Production*, 19, 667–679.

Chand, S., Davis, J. F. 2010. What is smart manufacturing. *Time Magazine Wrapper*, 28–33.

Chen, X. 2018. A framework for energy monitoring of machining workshops based on IoT. *51st CIRP Conference on Manufacturing Systems* (pp. 1386–1391). China: Procedia CIRP.

Duflou, J. S. 2012. Towards energy and resource efficient manufacturing: A processes and systems approach. *CIRP Annals*, 61, 587–609.

Eleni, Z. S. 2014. Towards a framework for energy aware information systems in manufacturing. *Computers in Industry*, 65, 419–433.

Fotopoulou, E., Zafeiropoulos, A., Terroso-Sáenz, F., Şimşek, U., González-Vidal, A., Tsiolis, G., Gouvas, P. 2017. Providing personalized energy management and awareness services for energy efficiency in smart buildings. *Sensor*, 17.

Franke, J., Charoy, F., El Khoury, P. 2013. Framework for coordination of activities in dynamic situation. *Enterprise IS*, 7(1), 33–60.

Garetti, M. T. 2012. Sustainable manufacturing: Trends and research challenges. *Production Planning & Control*, 16, 83–104.

Gordic, D. B. 2010. Development of energy management system – case study of Serbian car manufacturer. *Energy Conversion and Management*, 51, 2783–2790.

Groot, D. 2001. Energy saving by firms: Decision making, barriers and policies. *Energy Economics*, 23, 717–740.

Haller, S. K. 2009. The Internet of Things in an enterprise. *First Future Internet Symposium* (pp. 14–28). Vienna, Austria: Springer-Verlag.

Kannan, R. B. 2003. Energy management practices in SME – case study of a bakery in Germany. *Energy Conversion and Management*, 44, 945–959.

Menz, B. A. 2015. Die ETA-Fabrik. Implementierung eines Energiemonitoringsystems. *Fachkonferenz Die*, 8.

Müller, E. L. 2009. Improving energy efficiency in manufacturing plants case studies and guidelines. *16th CIRP International Conference on Life Cycle* (pp. 465–471). Cairo, Egypt.

Ramakrishnan, R., Gaur, L. 2016a. Feasibility and efficacy of BLE beacon IoT devices in inventory management at the shop floor. *International Journal of Electrical and Computer Engineering*, 6(5), 2362–2368.

Ramakrishnan, R., Gaur, L. 2016b. Smart electricity distribution in residential areas: Internet of Things (IoT) based advanced metering infrastructure and cloud analytics. *2016 International Conference on Internet of Things and Application* (pp. 46–51). Pune: IEEE.

Ramakrishnan, R., Gaur, L. 2016c. Application of Internet of Things (IoT) for smart process manufacturing in Indian packaging industry. *Information Systems Design and Intelligent Applications*, 435, 339–346.

Rohdin, P., Thollander, P., Solding, P. 2006. Barriers to and driving forces for energy efficiency in the non-energy intensive manufacturing industry in Sweden. *Energy*, 31, 1836–1844.

Scipper, M. 2009. Energy Related Carbon Dioxide Emission in US Manufacturing.

Vijayaraghavan, A., Dornfeld, D. 2010. Automated energy monitoring of machine tools. *CIRP Annals Manufacturing Technology*, 59, 21–24.

Watson, R. T., Boudreau, M. -C., Chen, A. J. 2010. Information systems and environmentally sustainable development: Energy informatics and new directions for the IS community. *MIS Quaterly*, 34, 23.

Weinert, N., Chiotellis, S., Seliger, G. 2011. Methodology for planning and operating energy-efficient production systems. *CIRP Annals Manufacturing Technology*, 60, 41–44, doi: 10.1016/j.cirp.2011.03.015.

Bibliography

Bi, Z. M. 2008. Development of reconfigurable machines. *The International Journal of Advanced Manufacturing Technology*, 39, 1227–1251.

Bi, Z. M. 2011. Revisit system architecture for sustainable manufacturing. *J-SustaiN*, 3(9), 1323–1340.

Emil, J. 2005. A wireless body area network of intelligent motion sensors for computer assisted physical rehabilitation. *Journal of NeuroEngineering and Rehabilitation*, 2–6, doi: 10.1186/1743-0003-2-6.

Güleryüz, G. B. 2016. Multi criteria group decision making approach for smart phone selection using intutionistic fuzzy TOPIS. *International Journal of Computational*, 9, 709–725.

Janak, L., Hadas, Z. 2015. Machine tool health and usage monitoring system: An intitialinitial analyses. *MM Science Journal*, 794–798, doi: 10.17973/MMSJ.2015_12_201564.

Kagermann, H., Wahlster, W., Helbig, J. 2013. *Umsetzungsempfehlungen für das Zukunftsprojekt Industrie 4.0: Deutschlands Zukunft als Produktionsstandort sichern*. Frankfurt: Geschäftsstelle der Plattform Industrie 4.0.

Marzband, M., Parhizi, N., Savaghebi, M., Guerrero, J. M. 2016. Distributed smart decision-making for a multimicrogrid system based on a hierarchical interactive architecture. *IEEE Transactions on Energy Conversion*, 31, 637–648.

Nishioka, Y. 2015. *Industrial Value Chain Initiative for Smart Manufacturing*. Tokyo, Japan.

Park, J. 2015. *Korea Smart Factory Program*. Tokyo, Japan.

Paul, J. A. 2015. Bottom-up approach based on Internet of Things for order fulfillment in a collaborative warehousing environment. *International Journal of Production Economics*, 159, 29–40.

Riddick, E. W. 2013. *Panel on Enabling Smart Manufacturing*. USA: State College.

Rittinghouse, J. W., Ransome, J. F. 2016. *Cloud Computing: Implementation, Management, and Security*. CRC Press, Boca Raton.

Rockstorm, J. 2018. *The Fourth Industrial Revolution: What Is Means, How to Respond*. World Economic Forum.

Sanjay, J. N. 2001. Virtual factory: An integrated approach to manufacturing systems modeling. *International Journal of Operations & Production Management*, 21, 594–608.

Scheer, A. C. 1992. *Architecture of Integrated Information Systems: Foundations of Enterprise Modelling*. Berlin–Heidelberg: Springer-Verlag.

SMLC. 2011. *Smart Manufacturing Leadership Coalition, Implementing 21st Century Smart Manufacturing*. Washington DC: Workshop Summary Report.

Stich, V. N. 2015. Cyber physical production control: Transparency and high resolution in production control. *IFIP Advances in Information and Communication Technology*, 459, 308–315.

Zhang, Y. F., Zhang, G., Wang, J.-Q., Sun, S. 2015. Real-time information capturing and integration framework of the internet of manufacturing things. *International Journal of Computer Integrated Manufacturing*, 28, 811–822.

Zhong, R. Y. 2013. RFID-enabled real-time manufacturing execution system for mass-customization production. *Robotics and Computer-Integrated Manufacturing*, 29, 283–292.

7

Logistics Optimization

7.1 Introduction

Logistical systems form a key link between industrial production and market circulation on the one hand while also helping with people logistics involving transportation such as air, water, land, and rail on the other. Logistics management circulates around three aspects: the flow of material and people, the flow of information, and, most importantly, the time taken to service a supply request. While requirements are for a point-to-point pick and delivery, this involves multimodes of logistics and often different service providers to coordinate to minimize costs, time, and ensure fastest possible delivery often involving geoboundaries and multiple currencies. Of late, reverse logistics (Hawks, 2006), which deals with reuse of products and its constituents, including remanufacturing and refurbishing (Anderson, 2005), has gained traction due to the focus on green operations, where the resource goes back by at-least one or multiple steps in the supply chain. Logistical focus has been the right product at the right time and place at the right price in the right condition, to the right customer (Mallik, 2010) and with the IoT, digital connected objects are becoming the focus point. Intelligent products need to have a unique identity, be capable of two-way communication, that is, both send and receive instructions and if possible act on it, store data about itself, is autonomous and capable of decision, and can broadcast information about itself to nearby connecting objects as a service (Wong, 2002). Production logistics can be either inbound relating to purchases and from suppliers to manufacturing lines, or outbound storage and movement from production line to end user. Logistics may take either of two forms, one that is material focused and the other that is resource focused. Logistical nodes may either be factories and assembly lines, depots or warehouses of storage, distribution centers, transit or drop points, and retail stores or point of sales and complexity increases with the number of drop points (zero as in direct supplier to customer deliveries or one level with a central warehouse or multilevel with many warehouses). Even within manufacturing facilities and warehouses, merchandising stock in a storage location is also a function of logistics and sometimes this function gets outsourced involving third-party logistics. Traditionally, automation of logistics involves using computer software and hardware to improve efficiency of the function using RFID or barcode technology. At the country level, the World Bank has designed a Logistics Performance Index (LPI) (World Bank, 2017), to enable companies identify and benchmark the efficiency of their trade logistics. The six dimensions of these are clearance by border control, quality of logistics infrastructure, shipment pricing competitiveness, quality of logistical services, consignments track and trace, and finally timeliness of shipments.

The use of the IoT and connected digital objects can plug issues such as transportation delays, cargo monitoring, IT point of failures, and in case of perishables, the loss of quality with every delay. The next generation Logistics 4.0 solution will use the IoT for real-time automated solutions.

7.2 Challenges in Logistics

Logistics is a subset of the supply chain, which also comprises of distribution, network planning, and supply chain process development (Figure 7.1).

The supply chain comprises of all activities (and the corresponding entities owning these activities) starting from the procurement of raw material (which may be finished goods for some other firm) to shipping final goods to the end consumer and then recycling it once the useful shelf life has been completed (Beech, 1998). Organizations near the source are referred to as "upstream" while the ones near the end customer are referred to as downstream. A manufacturing firm can extend its capabilities in both directions using forward and back integration strategies. Logistics deals with ensuing the uninterrupted flow of material and information, synchronously. In the case of discrete manufacturing complex machinery, this may involve thousands of independent parts to assemble in a defined bill of material. Unless the logistics function is efficient, it may lead to holding of excess and unwanted inventory, blocking capital and revenue streams, and ultimately slowing down all components of the economy. Each organization has conflicting demands. To be able to ship items on yesterday's basis at lower costs and ensure good

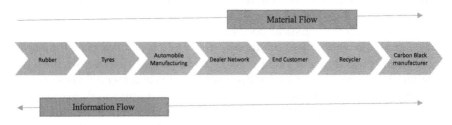

FIGURE 7.1
From rubber to carbon supply chain.

quality, items comprise of both physical aspects as well as services that are intangible but very crucial for the product to function. Logistics functions to ensure both of them reach the market as expected. Similarly, new product development is also aided by logistics by making products available at low costs. Quality issues are another area where logistics can efficiency play a role in making sure that no defective or short shipments takes place and in case of items like perishables, which require a controlled environment to transport, the same is maintained from start to finish. Customer waiting time or acceptable lead time is directly influenced by logistics, since a delay could cost a customer order, or could further impact their operations leading to a cascading delay effect along the supply chain ecosystem. Cost reduction is another core challenge of logistics that directly results in an aggregation of costs further downstream. Bulk manufacturing products with low margins are primarily driven by cost competitiveness. Logistics also provides support functions to bring in product differentiation to make it more attractive to the customer such as waiting time, controlling variance, and aligning deliveries as per plan or expectations. For example, a manufacturing firm in India came up with the slogan, "We delay we pay," where any material shipment that was later than 72 hours for an all India delivery was provided absolutely free. However, this was possible only by the adoption of advanced technologies including GPS tracking so that real-time tracking of material was possible on a continuous basis.

7.3 Logistics Costs

An end customer can derive value from a product or service if they can get a better return from the product than any other similar competing product of comparable costs with similar risks. Costs can be incurred on account of inventory, transportation, working capital, and cost of finance and may be fixed costs or variable costs and similarly direct or indirect costs. Logistics can help reduce costs by ensuring to be the first to hit the market, reduce inventory holding costs by supporting concepts such as just-in-time (JIT) and other lean inventory models, reduce transportation time and costs by providing optimized options, and reduce production costs by ensuring close to zero idle time. Costs ratios that can be directly assigned to inefficient logistics include wrong shipment ratio, penalties due to delays, inbound material availability, and disruption costs due to material shortage, and in-store nonavailability. Logistics costs can be divided as per the following approaches:

- Fixed and variable costs based on the activity volume, while examples of the former include warehouse rental, leased vehicles fixed rental, cost of electrification of warehouses, and fixed cost software.

The latter includes fuel expenses on transport, direct material costs, labor costs for material movement, and software costs on a use and pay service model. Fixed costs in logistics such as warehousing costs cannot be altered in response to falling sales volumes.

- Direct and indirect costs are based on the extent to which a specific cost can be apportioned or allocated to a certain cost head. Direct labor and material costs used in logistics management comprise of the first one, while costs such as transportation are indirect costs since they may be used for multiple products and activities and it may be difficult to exactly identify the units against each cost.

- Costs may also be engineered or discretionary depending on the ease with which they can be allocated and identified. In the former, there is clear input-output relation, while in latter, the relation is not clear (Dale, 1995). The costs can be converted from discretionary to engineer by dividing costs into preventing error costs, detecting error costs, and internal and external failure costs.

7.4 Autonomous Logistics

Intelligent products have revolutionized the field of product life cycle management and this section will focus on autonomous logistics and its requirements including process changes required. Conventional planning methods have serious limitations in a complex logistics network. The aim is to develop decentralized, hierarchical, planning and control capabilities in digital objects as part of autonomous cooperating logistics. Usually by applying new concepts, new properties of a system named "emergence" come up (Farideh, 2017), which alter the fundamental structure of organizations. Logistical "smart things" have to be characterized by their unique identification, communication, and collaboration, equipped with sensors that can detect information and readings about self and surroundings, must have local storage, must be capable of autonomous operations, and must have user interface (McFarlane, 2003). Other similar definitions of intelligent objects exist in literature explaining their use in supply chain optimization (Wong, 2002). For autonomous logistics, involving hundreds of complex objects, cooperation and interaction are needed. The common definitions are (Hulsmann, 2007):

> Autonomous Control describes processes of decentralized decision-making in hierarchical structures. It presumes interacting elements in non-deterministic systems, which possess the capability and possibility to render decisions.

One of the key identifiers of this decision is the concept of emergence of new structures aided by technology, which will aid the logistics function and its end objectives (Ueda, 2004). A typical logistics business function would require interaction of multiple agents (i.e., a MAS) where each of the agents represent an actor in the logistics system such as products, machines, assembly lines, robotic arms, packaging machines, inventory silos, and others. Each of these agents can coordinate and communicate passing instructions to each other in advance for meeting the next task. The first requirement is of hardware or embedded sensors in each of these agents that can access state, and present the context and environment data, the detailed aspects of which are mentioned in existing literature (Hans et al., 2008). The next requirement is of having a unique identifier that can be enforced using concepts like RFID, Bluetooth beacons, NFC tags, amongst others, and which can be attached to metal or plastic parts and can transfer signals that can be picked up by readers within a specified range. While this is workable in indoor conditions, the external communication may require concepts like NB-IoT and LTE-IoT along with GPS to provide the connectivity. Agents must also be equipped with decision-support software systems and algorithms to provide information on job scheduling, inventory movement, and machine assignments. Another requirement is of conflict resolution when there is a breakdown of machine and the requirement is to resolve this like human operations. In such a case, the agent must signal its inability to process work further and all pending transactions need to be reassigned to other working machines. This conflict becomes critical when the agent that suffered a breakdown is unable to communicate its last working state to other agents. Here, a manual workflow may be required or systems might be programmed to have all agents also communicate their heartbeat or functioning state at regular intervals and in case a heartbeat is not received in a stipulated interval, the system might proceed assuming a breakdown and reschedule operations on the defaulting node. Recent technology advancements in active RFID solutions have increased usage in merchandise tracking at every step of the supply chain process using a process called Path Authentication to ensure trustworthy and reliable delivery. Using Path Authentication, the entire travel path of the merchandise can be audited and tracked to ensure it missed no mandatory steps in the process such as quality inspection, activities required for export compliances like fumigation, and others.

7.5 The IoT-Enabled Activity-Based Costing

Traditional methods of allocating indirect costs by dividing them on the basis of direct costs is difficult in a complex logistics operation. Also, the overheads as compared to direct labor costs are multiple times with product diversification and large-scale manufacturing (Cooper, 1988). Activity-based costing eliminates this issue by recognizing that tasks lead to costs, and breaking down the logistics process into subtasks such as picking, loading, transport, and delivery is very useful for logistics cost measurements and managers can utilize this well since they have visibility on tasks (van Damme, 1999). With adoption of the IoT, it is possible for each task to be monitored using connected digital objects, for example, loading can be measured by RFID tags embedded in objects to be shipped. A manufacturing plant of flexible packaging film line production in India operates over five production lines each with an annual 18,000 tons production capacity. Each line can manufacture multiple products with a different combination of raw materials and chemical or metallized treatments. This requires changeover depending on the batch run size and as there are machine breakdowns, the firm wanted to adopt an activity-based cost apportionment on each line. The IoT sensors were attached to each line and used to convey the running time and idle time when the lines were put on mechanical maintenance mode by activating an actuating switch. Based on this, the exact cost could be assigned to each maintenance activity. Further cost information can be combined with time information, to create value (Bicheno, 2004). Previous literature has linked performance measurement systems to supply chain practices (Brewer & Speh, 2000) and also listed a set of characteristics for cross-supply chain measures (Derocher, 2000) namely fulfilling customer orders on time, fulfilling supplier receipts on time, internal defect rates, rate of introduction of new products, reduction in costs, goods flowrate, lead time from order to delivery, and financial flexibility. Many of these can be automated using the IoT functions such as material tracking systems powered by RFID or Bluetooth Low Energy (BLE) using location-based services and information capture.

7.6 The IoT and Inventory Consolidation

Consolidation of decentralized inventory can help firms save costs by avoiding duplication and maintaining low safety levels, but proper coordination of the supply chain must be ensured to avoid shortages and offset of benefits due to higher transportation costs. The IoT-enabled racks and bins help detect stock levels by measuring weight or count of material stock and passing information on as soon as it reaches some defined threshold that serves as a triggering point. Centralized inventories are more relevant where costs of holding inventory is much higher than the distribution costs such as semiconductor devices. It also adds the benefit of having full container load dispatches; however, in case the service time window is shorter, this model might prove counterproductive. With the IoT environment, firms can experience the best of both worlds by having a real-time communication between central warehouses and small decentralized stores. The IoT-automated measuring equipment fitted to devices can also help predict usage or consumption patterns making it easier to have a real-time assessment on the quantity to be shipped from the central warehouse. This also provides the ability to manage region specific volatility from the central warehouse. Nike has a centralized warehouse in Laakdal, Belgium, which is spread over 200,000 square meters and caters for 45,000 customers. The benefits include single-point destination for all supplier inbound materials and holding inventory and alignment of staff as per demand allowing for dynamic scaling up and down.

7.7 The IoT and Consigned Inventory

Some business models require supplier inventory to be placed at customer premises, where the inventory is owned for a specific period by the supplier. The customer may withdraw material for usage and is billed on actual usage; however, at the end of the period, the inventory has to be compulsorily purchased by the customer, the used stock being replenished by the supplier on a need basis. In such a scenario where the store might be unmanned, the IoT sensors fitted inside these material stock and the IoT readers fitted at the entrance and exit as part of a smart warehouse management system (Ding, 2013), can help account for material movement in and out of the facility, (thereby denoting a consumption or replenishment) with user intervention. This information can flow back to the supplier through connected devices, Wi-Fi or LTE, and hence generate automatic billing. Such a setup can also help in ensuring that the oldest stock is used up first to prevent losses due to expiration or damage.

7.8 Case Study: Industrial Logistics

Cost pressures on manufacturing industries are enormous and logistics form a key component of the unrealized profits and also costs. An Indian company "X" into core manufacturing of engineered product filling machines, required end-to-end visibility of its inventory and also transmission. The company had global manufacturing locations and huge warehouses to store raw materials, semi-finished, and finished goods. The company wanted to ensure process discipline while also having end-to-end inventory traceability since they were perishable, could be mistakenly intermixed since they had poor visible differentiation, but had a lot of variants on account of internal process treatments. This often led to costly customer returns when wrong shipments were delivered including penal charges, litigation, and logistics transport fares—most of which were intercountry water shipments. "X" decided to use the IoT solution for the same and identified active RFID as one of the technologies, with key constraints identified as path authentication to be used but with algorithms supporting minimum computational power, having real-time and continuous access to all concerned supply chain entities (Cai, 2012). However, in the case of warehouses where the area is huge and inventory items are stacked, direct line of sight for RFID was difficult to achieve. Hence, another hybrid technology, BLE beacons were deployed, which unlike RFID, did not have a line of sight requirement but had a major limitation of varying signal strength on one hand, which made pinpointing distances difficult, and also the range was higher, thereby functions requiring proximity solutions faced issues. The major business challenges to be addressed were inventory control and tracking, stock taking or cycle counting on a real-time basis, tracing efficiency of the supply chain logistics function including time spent in different stages of the process plant in terms of batch operations, and identifying potential bottlenecks at each of the stages. Also, since the complete process had many steps, the defects as a matter of practice were attributed to the preceding stage, However, this approach was adopted more due to convenience and was prone to judgmental errors. It was required to identify and correlate data, to identify the exact point of defect. Further operational efficiencies could be introduced by having faster detection of items in the warehouse ready to be shipped. The technology involved setting up requisite hardware, provisioning mobile devices in the form of tablets, provisioning integrated software that could do Path Authentication, use Cryptography Digital Ledgers based on blockchain for auditability, and use evaluated RFID tags and BLE beacons for a case-to-case basis depending on the respective strengths as highlighted in Table 7.1. RFID-active tags emit a serial number and optionally some additional data, while passive tags reflect the signals received from an antenna back to the antenna. Active tags can also have sensors to send information about the object state and surroundings like temperature. BLE beacons are hardware transmitters that can transmit their identifier information to any receiving Bluetooth device. RFID tags are capable of maintaining their arrival time within a zone and their exit time, hence allowing its tracking, while BLE beacons can just identify if a tagged object is within a zone or not. BLE beacons have been used for providing people with contextual information in airports (O'Connor, 2014; Perin, 2017).

The technology architecture involved selecting either active RFID tags of a type of active ultra-wideband, and BLE stickers or a mix of both, using a big database such as MongoDB for storing the data preferably on a cloud service and using rich analytics for measurement and reporting. The following benefits were observed as a result of the implementation (Ramakrishnan & Gaur, 2016).

- Faster time to locate inventory from warehouses, which was assessed using a before and after scenario in terms of time taken from the decision to ship a specific item to the time to exactly locate it in the warehouse.

TABLE 7.1

Comparison of RFID and BLE

Aspect	RFID Active	BLE
Range	Near Site	Over 100 meters radius
Accuracy	Nearly Exact. Read accuracy, which is the accuracy of reading each and every tag, is almost 100% in active tags, while passive tags may suffer from interference or blockage due to metals. Active ultra-wideband tags have a location accuracy of centimeters	Approximation correct to certain meter distance since the Bluetooth signal strength fluctuates by itself and also based on interferences
Coverage	Covered limited areas due to line of sight requirements	Could coverage huge warehouses
Costs: "item to trace" level	Price of active tags was as low as 2 USD and could be reused; however, the cost of an RFID reader/writer was expensive at 2,000 USD per unit	Beacons cost a lot of money per unit as much as 50 USD, and could be reused; however, the cost of more simple readers, such as a Raspberry-PI-based small-form factor computing, or mobile tablets, was lower at 60–100 USD
Observations	RFID was good for pinpointing location	BLE beacons were good for real-time inventory cycle count

- Higher accuracy of tracing items. Earlier process used to be prone to user mistakes but using this unique identification and tracking mechanism, it was easier to exactly identify with significantly higher accuracy.

- Process improvement steps could be taken based on the navigation path from start to finish as time taken at different stages could be easily measured.

- Real-time stock cycle count could be established as all digitally empowered objects were continuously tracking information that could be traced across and detected.

- With GPS sensors inbuilt into the beacon, exact movement of the materials even outside the plant could be tracked till the end destination, thereby allowing customers to have real-time visibility on their consignment arrival status.

- End-to-end tracking of inventory enabling prevention of loss and damages to the goods.

- Vision picking and scanner-based pick-operation by allowing items to be located and clubbed together including similar lots and based on serial numbers.

- Rich data analytics were possible on inventory movements providing insights into the warehouse operations.

- Using robotic arms that could coordinate wirelessly with the beacons, it was possible to transmit information on special packing instructions, which could be executed separately.
- The beacon readers could be integrated into moving drones, which would enable performing trace operations during nonworking hours.

Overall, the company managed to save a lot in terms of cost reduction, better optimized inventory schedules, streamlined operations, reduced time to ship out products, providing full transparency to the end customer, ensuring warehouse optimization through real-time assessment of inventory requirements, and thereby tightly integrated different warehouses.

7.9 Conclusion

Manufacturers, retailers, distribution lines, shipping, and planning agencies all require to manage the flow of raw materials, semi-finished, and finished products through the supply chain. Loss due to pilferage, damage, wrong shipments, and returns due to time delays are a major contribution to financial revenue losses. The complexity of the supply chain and logistics function is high in large organizations and very diverse and hence use of real-time location systems is necessary to improve logistics function. With the IoT technology, real-time location services through integration of appropriate sensors is relative easily, along with technologies like NB-IoT and LTE-IoT and GPS and GLONASS for external location tracking. The IoT passes information seamlessly between core systems and enables digitization of physical products. This helps with asset and inventory tracking, increasing efficiency, improving revenue opportunities, and bringing operational effectiveness. Further machines like conveyor belts and robotic arms can easily exchange information, helping in a streamlined machine-to-machine coordination. It can also help bring out structural changes in logistics operations of a firm by moving toward centralized warehouses, setting service levels for customer order fulfillment, and enabling automated unmanned warehouse operations. This chapter discusses how many facets of the IoT technology are available and each of them may have different levels of utilities based on the scenario under consideration. Organizations can adopt the IoT to bridge the gap between physical movement and digital movement of goods. The IoT miniature devices attached to logistics items, cargo, and transport fleet can transform them into smart parcels or smart transportation with the ability to broadcast contextual information about their locations. This enables users and stakeholders to monitor logistical movements in a centralized console to identify and track bottlenecks and even handle unexpected points of faults and failure by dynamically adopting processes.

References

Anderson, H. A. 2005. The consumer's changing role: The case of recycling. *Management of Environmental Quality: An International Journal*, 16, 77–86.

Beech, J. 1998. The supply-demand nexus. In *Strategic Supply Chain Alignment* (pp. 92–103).

Bicheno, J. 2004. *The 'New' Lean Toolbox*. Buckingham: Picsie Books.

Brewer, P. C., Speh, T. 2000. Using the balanced scorecard to measure supply chain performance. *Journal of Business Logistics*, 21, 75–93.

Cai, S. D. 2012. A new framework for privacy of RFID path authentication. *10th Intnl. Conference on Applied Cryptography and Network Security*. Singapore.

Cooper, R. A. 1988. Measure costs right: Make the right decisions. *Harvard Business Review*, 96–105.

Dale, B. A. 1995. *Quality Costing*. London: Chapman & Hall.

Derocher, R. A. 2000. Six supply chain lessons for the new millennium. *Supply Chain Management Review*, 34–40.

Ding, W. 2013. Study of smart warehouse management system based on the IOT. *Intelligence Computation and Evolutionary Computation*, 180, 203–207.

Farideh, G. 2017. *Bringing Agents into Application: Intelligent Products in Autonomous Logistics*. Bremen, Germany.

Hans, C., Hribernik, K. A., Thoben, K.-D. 2008. An approach for the integration of data within complex logistics systems. Dynamics in Logistics. *First International Conference. LDIC 2007* (pp. 381–389). Springer: Heidelberg.

Hawks, K. 2006. What is reverse logistics? *Reverse Logistics Magazine*.

Hulsmann, M. A. 2007. Understanding autonomous cooperation & control: The impact of autonomy on management. *Information, Communication, and Material Flow*. Berlin: Springer.

Mallik, S. 2010. Customer service in supply chain management. In H. Bidgoil (Ed.), *The Handbook of Technology Management: Supply Chain Management, Marketing and Advertising, and Global Management*. New Jersey: John Wiley & Son.

McFarlane, D. S. 2003. Auto id systems and intelligent manufacturing control. *Engineering Applications of Artificial Intelligence*, 365–376.

O'Connor, M. C. 2014. Major Beacon Deployment Takes Off at Miami International Airport. Retrieved from RFID Journal, September 24, 2018, https://www.rfidjournal.com/articles/view?12205.

Perin, E. 2017. Brazilian Government Announces Internet of Things Study. Retrieved from RFID Journal, October 9, 2017, https://www.rfidjournal.com/articles/view?16703.

Ramakrishnan, R., Gaur, L. 2016. Feasibility and efficacy of BLE beacon IoT devices in inventory management at the shop floor. *International Journal of Electrical and Computer Engineering*, 6, 2362–2368.

Ueda, K. A. 2004. Emergent synthesis approaches to control and planning in make to order manufacturing environments. *Annals of the CIRP*, 53, 385–388.

van Damme, D. A. 1999. Activity based costing and decision support. *International Journal of Logistics Management*, 10, 71–82.

Wong, C. Y. 2002. The intelligent product driven supply chain. *Proceedings of IEEE International Conference on Systems, Man and Cybernetics*. Tunisia: IEEE.

World Bank. 2017. http://lpi.worldbank.org/

8

Distribution Channel Management

8.1 Introduction

Distribution channels include sales channels that can influence buyers and share new product information, delivery channels for physical product delivery, and service channels to provide aftersales services. The distribution channel members include a firm's sales function team, carrying and forwarding agents or carrying and selling agents, distributors and wholesalers, dealers' resellers, and stockist and agents and brokers operating of a brick and mortar shop or electronically, all of whom have different roles and contracts with the manufacturer (Figure 8.1).

The patterns of distribution predict the intensity and the service levels and may be intensive distribution, selective distribution, or exclusive distribution. The first focuses on covering every retail outlet and ensuring maximum reach and is needed for industries like pharmaceuticals and automobile spare parts. The second has multiple outlets and each of the outlets is designed as per brand requirements to have identity and these are used in jewelry or electronic goods such as Apple products. The last one focuses on very selective outlets and may be exclusive for one product and might be company owned like Ford cars or Titan watches. They usually have the maximum control option available with the products team. A distribution strategy needs to incorporate customer service levels, network structure, and key performance indicators, some of which are very difficult to measure and record and are also industry dependent. There are external environmental challenges, also more so in industries that are dynamic and unpredictable (Sahadev, 2004). The distribution activities include dispatch planning, beat plans by sales representatives in a territory and associated market coverage, and travel planning, some of which may be outsourced and hence require close monitoring. Another KPI for distribution efficiency is zero complaints from

FIGURE 8.1
A consumer distribution chain.

customer, and the ability to handle demand spikes, minimizing stock-out situations, and minimizing product damages. Distribution channels help in resolving a number of discrepancies arising due to the different places where production and consumption take place. The first being spatial discrepancy, that is, the distance between producer and consumer. The second is temporal discrepancy, which is the time lag between production and consumption. The third is quantity discrepancy, that is, usually a large quantity is produced but each customer consumes a very small quantity, bringing all goods of the same company or different brands together in one place to facilitate the customer to make a choice, and also provide financial support due to the different payment modes and credit period. Some products follow a shorter product life cycle while others have a long cycle; however, at each stage, information needs to be captured to further pass on the information to the distribution channel (Ryans & Shanklin, 1984).

8.2 The IoT Applications in Retail Distribution

The IoT can help in both forward flow from company to customers and backward flow from customers to the company. It also facilitates the flow of physical material, title flow, and information flow and becomes useful in international marketing and where just-in-time (JIT) concepts get implemented (Rosenau, 1988). Retail is about attracting customers and measuring audience engagement in real time to provide insights about changing trends and customize content according to individuals. Display technologies like digital signboards, video walls, and advanced recognition mirrors are becoming part of everyday shopping experience. The economics of digital signage is entirely based upon audience reach and influence and also the ability to attribute this reach and influence to a favorable outcome such as brand or product awareness or even an actual sale. The IoT technology can help in this powering smart signage with full sensor integration, integrating with data sources to capture and provide customer feedback covertly using their facial expression identification, or giving customers a touch screen to enter their preferences. Supply chain modeling by way of simulations help in detailed analysis and a pragmatic approach toward distribution chain design and management (Swaminathan et al., 1998). Technology is transforming retail by connecting the personalized experience of the past with the projection of customer's future buying trends. Research has also reiterated that information sharing helps in generating superior profits for distribution channels (Ding et al., 2011). As described earlier, a proper distribution channel including retail stores are defined based on data and projections, these include but are not limited to the store type, the target market to be positioned to, the optimal product mix, customer service channels, post-sale services, and the store's overall market positioning. Retail shops are spread across marketplaces, retail chains and retailers may be food retailers, soft-line retailers, grocery retailers, etc. To understand how the IoT can help retail business, we identify the components that determine the retail environment success. The most important aspect is ensuring customer delight (McKenna, 1991). With the IoT using location-based sensors and services, it has become easy to follow a customer and record their preferences continuously so that any change in preferences and perceptions can quickly be factored in design and product development. These sensors can be embedded in loyalty cards or entry passes with the customers or simply by providing accessories that can bind with the customer's cell phone. What a customer picks up at a garment store, how much time a shopper spends in which section of a shop, and the ability to propose alternate products intelligently based on their current selection are features that the IoT-based solutions have been able to provide select retailers across the globe. With simple consumer items becoming connected digitally, every experience is becoming a digital experience. Initially, retailers used to fear the smartphone as people would end up comparing prices with online deals inside the shop. This has given way to acceptance of the smartphone as a unique means to identify and connect with the customer. Location-based Bluetooth beacons are one way of tracking smartphones and communicating instant deals to them. Some departmental stores, notably Hudson's Bay, have implemented Apple's iBeacon technology for locational analysis, contextual information offload, and customer time tracking (Mohr & Nevin, 1990). They have also leveraged the marketing mobile app "Swirl" to provide customized messages and promotional alerts. The data so collected can help in shaping the distribution cycle end points, namely the retailers, and by providing layouts to maximize brand visibility and appeal, and finalize merchandising layout strategies. Another adoption of the IoT in retail is in Hugo Boss stores, which have positioned heat sensors to map customer movements and place high-priced merchandise accordingly. Smart mirrors have shown promising results in identifying issues covertly such as time taken to trace items inside store or choosing substitutes (Gaur et al., 2017). These also transfer data about some of the hidden aspects related to distribution management namely merchandising preferences using the IoT technology embedded sensors inside smart mirrors. Proximity driven checkout and payment systems based on the IoT sensors can help reduce operating costs of the distribution chain by reducing time spent on making current transactions while allowing full transparency and record keeping in all merchandise dealing. There are prototypes of automated payment mechanisms where a digital wallet can be automatically debited based on an item detected using IP cameras and intelligent object recognition and artificial intelligence. The IoT-enabled smart shopping carts with a digital grocery list can exactly locate and even navigate to respective areas in a big shopping mall using robotic wheels and Bluetooth Low Energy (BLE) beacons as identifiers of location. Smart price tags that can be changed based on footfalls or demand can be affixed to products to display changes to price dynamically and also discounts can be announced and implemented on the fly, creating some sort of demand-driven pricing. Smart digital tags can be read by customers who can use their smartphone or a wearable device to quickly scan an item and call up product information, reviews, or social media reviews. Energy efficient lighting that increases or decreases as per foot falls or crowds inside shops can help distributors and retailers conserve energy costs. Robots with touchscreens for browsing inventory can lead customers to desired products at different levels of the distribution chain. Warehouses and retail stores can be equipped with the IoT-enabled smart shelves that can automatically replenish inventory levels when it goes down based on weight sensors or unit counters. Perishable items like fruit juices can have the IoT smart tags that can inform when expiry date is nearing so that they can be in combination with smart tags subject to dynamic pricing. If the expiry date is nearing, the prices can be decreased and a higher discount given on these products. The IoT can help streamline the supply chain part of distribution management by using smart objects power to transmit information in real time about each stage of the supply chain operation and information about stages before and after the current stage. Data visualization technologies in tandem with the IoT can help customers and employees of each segment of the distribution chain to track inventory. Based on these inputs, retailers and distributors can adjust pricing in real time and pass on benefits including the one derived from correctly estimating and planning during the lead time to the next customer. This will bring about consistency of pricing and prevent inflated profits or losses, overall creating a healthy ecosystem-wide distribution management process. Another benefit to the distribution chain can be the creation of new business models and revenue streams in the form of services and information-based application, through the IoT-connected objects and shipments. Customer satisfaction levels are also bound to rise with connected objects. The IoT device information will help retailers or distributors drive more targeted offers to their immediate customers. In the pre-internet era, shopping for clothing used to be a different experience as customers preferred to see and wear anything they wanted to buy. This would imply browsing fashion magazines or newspapers, identify the clothing apparel, visiting different stores, trying out combinations, and then buying. All this was aided by availability of free time, lesser traffic and travel timings, and lack of other means. Today, with internet and smartphones, customers are able to locate dresses sitting at home, order online, do a trial in the comfort of their home, and finally return it if not satisfied without moving outside. The aid to this mindset change, which till a few years back was considered unthinkable, was the unavailability of free time, lots of traffic congestion and hence travel time, and finally climate changes making it more unpleasant to travel outside. With the IoT taking this information availability to a new level, it might be possible for objects to interact with each other and with humans in the form of advertising or broadcasting their presence and arrival automatically without a human driven "information pull." This is very important in the case of high-tech industries where innovation and technology evolves at a very fast rate (Curry & Kenney, 1999).

8.3 Field Research

An in-depth questionnaire interview was conducted with senior managers responsible for retail operations in five leading Indian wholesalers and retailers including multinationals, Indian origin and owned, and a local grocery chain with at least eight outlets in India. The interviews were based on a short discussion guided by the imperative to understand challenges in distribution and understand by proposing solutions based on the IoT how these challenges can be influenced. The following are the key takeaways:

- The first challenge faced in distribution management today is in moving from generalized- to personalized-based interaction with customers at the same point of time, that is, in a shop, while a person who is visiting for the first time may need to stand in a queue but a frequent shopper may get priority or a separate line for billing.
- Another challenge is in managing customer engagements and retention issues. Next in line are evolving Omni-channel strategies with so many connected devices.
- Finally, a big issue is in understanding changing preferences across generations or age groups or even city and villages.
- All the interviewees raised data security and privacy issues as a main concern area including location tracking or preference tracking and sharing its information with advertisers without user content. Also, a lack of regulations and legal frameworks is a gray area and managing stock and supply chain visibility is a big area where improvements would be fruitful.
- The distribution chain suffers stock outs and over stocking, both of which are related to the information flow discrepancy. Usually, information was shared based on rough assessments relating to predicting the demand. Also, minimum stock levels were maintained at each point of the distribution chain.

Specific sector-wise challenges as observed by the interview in retailers have been highlighted as follows.

8.3.1 Consumer White Goods

The digital era has hurt this section most as people try out and study the features in-person in stores and then order online with a guarantee that the service and quality is ensured by the manufacturer and hence online and offline does not make any difference and usually online stores are better priced due to their reduced operating expenses on rental shops or electricity consumption. Customers feel happy talking to specialists in stores to get answers to their queries, which online takes too much time for someone to respond to. However, they like online features such as customer feedback as well. Delivery of material online seems to be more difficult from a retailer prospective than delivery from the physical shop, which is usually situated nearby.

8.3.2 Super Mart Items

The advent of super marts, where all of the customer's needs can be met in one single place, has been a significant development over the last two decades. Many of these marts have a forward and reverse strategy in place for an end-to-end integration. This saves time and the same benefits are extended by online stores as well. However, online supports with a "memory" option of last purchases thereby reduces time further.

8.3.3 Food Items

Generally, preference is for offline where people can see, smell, or even taste the item unless they are packaged and processed items. The IoT sensors that show nutritional values or recipes that can be developed using these food ingredients or helping locate where the item is can play a key role. Also, many items are perishable and have a limited shelf life. The IoT sensors can help check optimum temperature during transport continuously. The IoT sensors can also help in suitable placement such as lots whose expiry dates are nearing by thereby helping in quality control while helping retailers with better waste management and regulatory compliances relating to nutrition values or expiry dates. For example, frozen goods have to remain at a specific temperature that, till now, gets recorded by hand. The IoT sensors can automate this mundane task. With the IoT, retailers can expect reduced energy costs in the food segment using smart freezers, which take into consideration ambient outside temperatures and alert against door open or falling/rising temperature.

8.3.4 Fashion

In this field, interviewees were looking at how to have a sustainable approach in terms of costs, flexibility, and pricing. Omni-channel initiatives allow fashion retailers to cover everyone from mass markets to community centric groups. With demand for capturing data, which will help in profiling, flexibility, and diversity, future fashion retailers are projected to be around a limited number of stores supported by digital channels.

8.3.5 Logistics

The IoT will bring in the era of experienced-based retailers and retailers are already implementing sensors and beacons for dealing with logistics, fleet management, and inventory management including item location tracking. Robots powered by the IoT are getting used for packaging and end consumer entertainment, for example, Italian grocery store Coop.

8.3.6 Energy Management

Retailers are using the IoT or willing to use it for energy management of deep freezers or lights or sound adjustments such as ones used in Kroger, USA. There, the IoT sensors measure temperature every 30 minutes in deep freezers and are able to alert store managers and facilities engineers when and where levels drop too low. Carrefour, France, has established intelligent lighting systems. Ocado has developed an in-house IoT warehousing solution automating a majority of firm's retail operations from order to fulfillment.

8.3.7 In-Store Technology

Interviewees felt that stores and in-store technologies are extremely important for a rich human shopping experience, which is unparalleled in online stores. These technologies include digital screens, touch signboards, smart mirrors, augmented reality, and virtual tables. Also, the ability to capture key decision moments where a shopper buys or decides to drop off and exit the shop are important. Stores have to enable shoppers to shop at the instant through RFID tags or NFC or other technologies. Online and social technologies and mobile applications play a key role in customer engagement across the entire distribution chain and should be integrated with the in-store experience by means of sharing data. Interviewees believed that new technologies powered by the IoT, such as wearable's inbuilt into loyalty cards or badges or wrist bands, hold a lot of potential. Vending machines in retail shops with intelligent kiosks can be manned remotely and also dynamically controlled including price points or product offerings depending on climate, etc.

Below are the IoT technology components interviewees were made aware of:

- *Beacons*: These can be used to deliver personalized promotions to customers based on their proximity within a store or proximity to shelfs or sections in big stores. This can help based on the time spent in each location, the demand, or interest levels for each product, which could be aggregated across the distribution link to get a better assessment of actual stock required, thereby reducing stock outs or over-stocking issues. Areas receiving lesser interests can go as a feedback to the manufacturer to reduce generation.

- *Smart Thermostats/Lighting*: These can be used to provide adequate lighting and temperature while improving energy usage in wholesaler and retailer outlets depending on the number of people in a specific section of the store. With cost savings, the benefits of energy management can be passed on to all the stakeholders.

- *Sensors*: To monitor the quality of perishable food items and plan inventory to meet the demand, sensors can be embedded in the form of smart tags. The advantages being that in case of products requiring moderated environmental conditions, the damage to products on account of temperature variance or high pressure can be avoided preventing losses to inventory.

- *Smart Tags*: For storing product information, tags can be used. These tags can again be detected by high-resolution cameras or proximity sensors, thereby allowing for a quick checkout and preventing pilferages. This is more relevant at wholesale, where mass movement of items need to be factored.

- *Advanced Analytics*: This is used to study the visitor flow and purchase patterns to identify fast moving products, high traffic areas, etc. This will help, for example, in placing premium products in high-traffic areas or to categorize customers based on their frequency of visit, interest, and purchases done, thereby providing personalized services to high-valued customers.

There was an added observation that integrating these services, identifying the right hardware vendor based on expertise in implementing solutions, and costs were the main challenges and it was also enquired if such technology had been floor tested in some location.

8.4 The IoT Case Studies in Retail

Levi Strauss, a retailer in garments, worked with Intel to implement an IoT solution starting from its San Francisco stores aimed primarily at avoiding inventory distortion in the form of inventory stock out, wastage, and distortion. It launched a technology trial using RFID tags on all items of the store, which was integrated with a cloud platform for reporting and analytics. RFID antennas continuously detect and aggregate this data to ensure the monitoring of stocks and their current levels enable stocktaking within seconds. The system provides alerts when stock runs low by continuously scanning for products. The store has implemented sensor capabilities with video analytics to analyze the way customers move in the store to help optimize layout planning and merchandise placement and stocking.

The following are the results of this IoT enablement of the stores:

- *Inventory Visibility*: Levi's team now gets 100% visibility of their inventory with almost no need for physical stock counting even at subcategory level such as pack sizes or colors. Customers are able to locate stock and there is no lost opportunity due to stock-out.

- *Cost Reduction*: Stores do not lock up capital in unwanted stocks. Moreover, with efficient lighting, energy costs have gone down. Thefts and pilferage of items is almost negligible due to better detection using RFID tags.
- *Customer Understanding*: The system records not only items sold but also items picked up, touched, or moved around and then not purchased. Hence, the shop is able to map demand locally and what the barrier is for a customer to complete a specific sale. This can help with repricing and also how many areas the customer actually traversed before ending their shopping trip.
- *Effective Item Placement*: Sales have risen due to exact placement of items and user's ability to trace it easily without seeking additional help. Also based on a common pattern, items can be so logically arranged.
- *New Revenue Models*: Information so gathered on inventory movement can help Levi package and bundle offerings, giving customer discounts on packaged sales.
- *Customer Experience*: Using smart mirrors, customers can try out virtual clothing in their own privacy and seek help when required thereby saving time in trying out clothes and searching for different color and size combinations. A study by the authors had shown the effectiveness of a smart mirror in garment stores (Gaur et al., 2017).

8.5 Costa Coffee

This coffee vending chain operates 1.7 million vending machines in over 7 countries and has seen a decline due to vending kiosks that are touch operated and have more engaging features like advertisements or news or digital signage. A revamp of the old vending machines was done by the IoT, enabling it with touchscreens, payment gateways, and the ability to respond to taste preferences by remote monitoring and management of these new digitally connected machines. The machines have multiple hardware components integrated by software component and support interactive display boards. From the vendor side, it allows remote management with online monitoring and diagnostics enabling preventive maintenance and diagnostics for new vending machine design and development, reliable monitoring of temperatures for hot and cold beverages, data for campaigns and new product launches, and consumer identification by Bluetooth connectivity or beacons. The connectivity happens through SIM-based technology. There are social media links for broadcasting user preferences and also as a digital marketing campaign. The benefits of the system have been summarized below:

- Real-time data transmission and analysis including when to stock and when to replenish items.
- Alerts against impending machine faults.
- Just-in-time delivery and over-the-air customization of products and offerings.
- Understanding of individual customer preferences and also of overall sentiments.
- Using the vending machines as a marketing tool and a machine observer.
- Using a video camera detection of purchase choices and the time taken to exercise choices, which becomes a scientific source for preference analysis.
- Ability to charge differential pricing because of differentiated value-added features.

8.6 Conclusion

Distribution chain participants are experimenting with newer ways and technologies to meet the broader objective of having the right product at the right place, right time, and right costs. However, the distribution network is often so complex that proper flow of information is restricted, leading to slippages and breakdown in the chain of tasks and material flow of goods. The Internet of Things with its potential to make an object digitally empowered, self-broadcasting, and ubiquitous, addresses the fundamental problem of timely and real-time information flow. In the last leg from retailers to end consumers, the selling channel may get modified by doing away with traditional methods of retail advertising such as newspaper or banner ads or pamphlets, which are mass marketing modes, and may soon get replaced with tailor-made IoT-enabled targeted marketing for masses. The IoT presents an opportunity for all players of the distribution chain to develop an improved ecosystem connecting physical and digital worlds, allowing bidirectional, real-time interaction with consumers using the omnipresent smartphone of the consumer as the focal point for identification. Location awareness, preference detection, real-time information capture on behavior trends, some aspects which could not be earlier measured, can now be converted into data using the IoT sensors. This can provide a solution to behavioral issues in managing distribution channels, particularly industries that are highly dynamic and less predictable. However, there are technical challenges, cultural challenges, and data ownership issues that have to be overcome for a successful adoption of the IoT in distribution business units worldwide.

References

Curry, J., Kenney, M. 1999. Corporate responses to rapid changes in the PC industry. *California Management Review*, 42, 8–36.

Ding, H., Guo, B., Liu, Z. 2011. Information sharing and profit allotment based on supply chain cooperation. *International Journal of Production Economics*, 133, 607–631.

Gaur, L., Singh, G., Ramakrishnan, R. 2017. Understanding consumer preferences using IoT SmartMirrors. *JST*, 25, 939–948.

McKenna, R. 1991. Marketing and selling technology products to mainstream customers. In G. Moore (Ed.), *Crossing the Chasm*. New York: Harper Collins.

Mohr, J., Nevin, J. 1990. Communication strategies in marketing channels: A theoretical perspective. *Journal of Marketing*, 36–51, doi: 10.2307/1251758.

Rosenau, M. 1988. Speeding your product to market. *Journal of Consumer Marketing*, 5, 23–33.

Ryans, J. K., Shanklin, W. L. 1984. Principles of high technology marketing. *Business Marketing*, 100–106.

Sahadev, S., Jayachandran, S. 2004. Managing the distribution channels for high technology products. *European Journal of Marketing*, 38, 121–149.

Swaminathan, J. M., Smith, S. F., Sadeh, N. M. 1998. Modeling supply chain dynamics: A multiagent approach. *Decision Sciences*, 29, 607–631.

9

Green Manufacturing

9.1 Introduction

The manufacturing sector contributes more than 31% of the total energy spending worldwide (EIA, 2010) and of late there is a rapid push toward sustainability in manufacturing (Daily & Huang, 2001). Green manufacturing efforts focus on saving energy by lowering energy consumption, improving efficiency of production and logistics functions, and improving overall consolidated management (Weinert, 2011).

The use of the IoT technologies can help with real-time energy monitoring by way of data acquisition using embedded sensors and thereby monitoring individual machines (Ramakrishnan, 2016), something that was earlier not possible as energy management was focused on the net total consumption and breakup into individual components was a less scientific process.

Previous literature points out a multitude of methods deployed to monitor energy, model new processes that are energy efficient, simulation and optimization, and use of decision support systems (Muller, 2007). Adoption of lean manufacturing can help in reducing waste generation without any significant impact on productivity (Michaloski, 2011), and machines can also be made ecofriendly (Reich-Weiser, 2010) by reducing the time spent in the active operating stage for each machine (Gutowski, 2006).

A previous study has also indicated that substantial energy is wasted during the higher energy requirements during startup stage, or idle waiting time (Park, 2009). Initially, production systems were designed only to produce material and there was no requirement of making them digitally connected, hence they did not have the requisite capabilities of storing data, connecting securely, or interacting with each other. The IoT enablement of these machines can help in storing, sharing, and analyzing data about the machine's operating conditions, which include energy parameters (Vijayaraghavan, 2010) during down time and idle time, running time, and startup (Bunse, 2011).

Provisioning of smart meters is another way in which the IoT can be used for measuring energy requirements of a running factory (Haller, 2009). The IoT technology has been put to use in a number of manufacturing operations ranging from control of inventory (Meyer, 2011), preventive maintenance, planning for material (Poon, 2011), and control of wastages at shop floor level, and machine level monitoring (Zhang, 2011).

9.2 Framework for Green Manufacturing

The following is a framework for adopting green manufacturing practices in any process or discrete manufacturing industry based on saving energy, optimizing machine run state, improving operational efficiency, efficient use of raw material and labor, and ensuring uptime. This model identifies the activities that are not green and tries to use the IoT technology to capture data at each point so that required metrics can be measured.

As part of the framework stage one, three dimensions are envisaged, the energy consumption, the machine operating state, and the material and labor movement. All machines and subsystems should be IoT-sensor enabled, each of which should be capable of capturing information relating to the machine's operating state with reference time frame detailing time spent in each stage. This will help identify which states are unproductive and how much energy is consumed.

Further with information of the product or batch composition, energy consumption variance as per the product and corresponding batch run can be analyzed. Material movement from raw material stores to process or assembly lines and then in-between different process and assembly lines and then to a work in progress store or finished goods stores for outbound logistics is an area where it is necessary to track inventory and machines to reduce idle or unproductive time; further, as a facet of green manufacturing, locational awareness and contextual information including aging is required so that stock is not lost as wastage of expired stock (Giovanni, 2013). This stage focuses on enabling data acquisition and capture on a real-time basis so that information is available on energy consumption not only at the plant level but individual machines, individual processes, and run state and product wise (Figure 9.1).

This way, energy consumption data or machine-run stage data or even material movement data becomes an important source for planning the production activities for each product batch run, thereby improving energy optimization measures.

The data so obtained is fed into the manufacturing execution system and by way of integration to a digital ERP source so that each product can come with a transparent data on the energy used in manufacturing in its labeling. Manufacturing costs have both fixed and variable costs: it may be energy costs, manpower and ancillary machine costs, or green manufacturing aims at reducing the costs and associated energy footprints by maximizing production per unit of fixed cost for which it is necessary to first measure the costs.

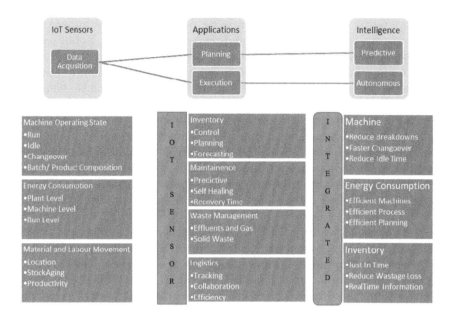

FIGURE 9.1
Facets of green manufacturing.

The above second level of the framework deals with improvements to the energy consumption by capturing the data in a relational database system that can be queried or the inputs channelized back for further production planning and scheduling. The last level deals with optimization based on intelligent patterns and autonomous behavior of machines where machines can go into idle state or power-saving state automatically with a decrease in workload sensed in real time based on accumulated data analytics. This also brings in lowering of costs, enables machine-to-machine (M2M) communication, and encourages a decentralized and, to an extent, autonomous decision-making process.

Another aspect of green manufacturing is the control of solid, liquid, and gas wastes generated in the production process. Manufacturing produces more than 50% of solid wastes in the world (Hill, 2004) and uses almost 20.2% of land water during the process (Gavronski, 2012).

The third aspect is of Green Training, since the human factor is an important influencer of green manufacturing driven top down from the leadership level (Bai, 2010). The green orientation, both internal and external, refers to the extent of emphasis on adopting green practices in a manufacturing concern and is reflected in its financial statements and annual reports (Banerjee, 2002). Numerous studies exist that show adoption of sustainable manufacturing can lead to competitive advantages in the long run by ensuring they can persist for the long term (Bai, 2010). Green innovation also refers to the ways a manufacturing firm innovates new methods and techniques unique to its operating business by minimizing unwanted effects and reducing environmental impact by way of avoiding rework or curbing pollution.

Last but not the least, is Green Supplier development that provides inputs that are manufactured using green or environmentally friendly methods since buyers need to get involved to train their suppliers (Krause, 1997).

The IoT sensors fitted at a supplier's place and controlled by the organization can help record the entire manufacturing process and ensure auditability of data in conformance to agreed contractual obligations. Similarly, energy meters at each level of distribution can help capture not only the manufacturing environmental impact but also the distribution cycle. Building green buildings as part of manufacturing facilities and using renewable sources of energy like solar and wind to cut down on normal fuel usage is another aspect of green manufacturing. Manufacturing facilities require the use of a number of ancillary equipment like chillers and boilers, which are very energy intensive, hence ensuring they are energy efficient can help reduce environmental impact. Daikin Applied, which has the world's largest heating, ventilating, and air conditioning systems, has adopted the IoT to measure the real-time energy consumption of its devices. Green practices such as recycling, energy efficiency, environmental emission and effluents control and compliances, preventing disasters, and reducing wastage due to expiry are some measures that can be implemented easily using the IoT. These not only reduce the extent of new product manufacturing but also ensure water bodies are not contaminated. Sensors reduce downtime, help track movement and disposal of materials, keep tab on recyclable material, and help in better optimization of logistics or transportation setup reducing carbon footprints. Real-time data can help in remote management and administration and also implement a Cradle-to-Cradle design (C2C), which follows a biomimetic approach to the design and development of products and provides social benefits.

9.3 Case Study

A leading global footwear organization adopted an ambitious plan to green its manufacturing process including assembly facilities and adopted a green orientation internally (toward its own facilities and activities) and externally (toward in extended value chain and ecosystem) in one of its popular brands of footwear.

The perceived benefits were customer satisfaction, especially for their global customers who were interested in seeing the energy and environmental impact of the finished product before buying it. The company hoped to have its packaging label reflect this information, which could also be auditable and accurate to the extent possible. It was understood that at every stage of the manufacturing process, even starting before the raw materials reached the warehouse, that is, at the supplier end itself, the process had to be tracked using data acquisition sensors and hence training the suppliers on the need of this was necessary.

There was a risk of losing some part of the profit but it was expected that environmental-focused and savvy customers would be attracted to the company and hence the sale units could pick up. Also, as part of the comprehensive strategy, significant reduction in inventory holding costs and energy costs were anticipated, which could make operations more profitable.

Another aspect considered was the option of recycling and take-back offerings, which could be recycled and provided as part of a corporate social responsibility to the needier sections of society.

The processing of raw materials, including sheep skin and leather, also has an environmental impact including usage of water. As a starting point, the IoT flow meters were used to estimate the water used for treatment of raw material and the water was recycled back to be used again and again. The waste so generated was compressed into solid waste, which could be used for civil work. Further analysis showed that almost 90% of environmental impact was as a result of material processing and production (Figure 9.2).

Materials management was streamlined by adopting the IoT-enabled active tags on all materials, be it raw material, work in progress goods, or finished goods. All machines in the assembly line, including the ones responsible for cutting the soles in required sizes, the ones used for creating the support structures for feet, and the ones integrating the soles with the supporting piece, were connected using M2M communication, whereby each machine could convey its output expected in advance to the next in the waiting assembly line, thereby preventing a machine idle state. Conveyor belts were equipped with robotic arms and all machines were fully automated and digitally connected. There were readers at regular intervals that were used to eliminate free hanging or unidentified stock. The machines that were involved in the production process are explained in Figure 9.3.

High precision and speed, the IoT-enabled digital cameras were used to weed out defects and could even stop the machine in case of frequent defects, signaling to the nearest human operator of the malfunctioning of the machine.

FIGURE 9.2
Footwear manufacturing stages.

FIGURE 9.3
Machines used in footwear manufacturing.

Each machine was equipped with energy meters that continuously recorded energy consumption of the machine along with details of the product batch run, which was getting stored in the ERP system and it was a key index to reduce the energy consumption for similar batch sizes and products at every subsequent run by at least a fraction of a percentage. This led to redefinition of key processes and also restructuring of the warehouse and other facilities to reduce man and material movements.

Further extending the reach to immediate customers and suppliers, energy-use labels were printed just at the wholesaler and exclusive retail outlets as energy efficient machines, lighting and ancillary equipment like chillers and boilers were used. Product promotions also involved the take back of old material from existing customers and the products were recycled to produce innovative community friendly offerings. The IoT readers installed at supplier premises also kept track of their raw materials used and their energy consumption at different stages.

The company incurred significant financial benefits as energy requirements came down multifold and sales revenue increased due to an optimized supply chain and better input prices due to green vendor initiatives carried out with the suppliers. With real-time data, the company was also able to implement full transparency by sharing data about its manufacturing footprint and energy requirements. In terms of compliances, the IoT-enabled emission detection meters were installed that could detect the kind of gaseous and liquid wastes being generated at each stage of the machine, although it was more during the raw material processing stages (Figure 9.4).

Also, there was a significant involvement of labor such as putting on adhesives and hence green training and focus was provided to the shop floor workers.

However, as a measure to check that they were following the process, an IoT-enabled camera with integrated software that could use image detection and analysis to check compliance was involved. This was able to identify any defects in the manufacturing process induced as a result of worker negligence.

The other contributors of environmental impact namely transportation and packaging and end of life wastes were addressed by having GPS-enabled systems for route planning and optimization and using the IoT for connecting different transportation modes.

Also, green manufacturing was integrated in the product design stage where energy and machine time was made an important role of the design strategy. All facilities were equipped with the IoT-enabled motion sensors that could switch on and off lights and air conditioning based on human proximity. At the end of the green manufacturing initiatives, the company was able to implement and monitor measures in the areas of energy use from lighting and machine operations, use of water and its recyclability to avoid spending potable water, emissions from chemicals such as adhesives and glues and colorings agents, and generation of solid waste.

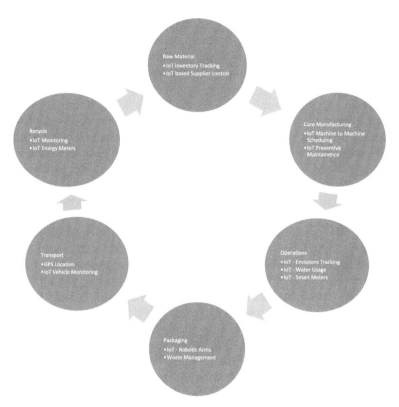

FIGURE 9.4
Application of the IoT in the footwear manufacturing process.

9.4 Collaborative IoT for Advanced Manufacturing

For overall green manufacturing ecosystems, firms have to collaborate and the IoT is appropriate to bring this, aided by cyber physical interactions. This can facilitate remote manufacturing, remote facilities operations, and reduce environmental impact of transportation. This also helps introduce agile strategies that can achieve engineering and manufacturing activities faster than traditional methods. Thus, impact of rework or wastage on account of customer requirement changes can be reduced. In a collaborative context, networking technologies like the cloud can be used to transmit information about physical activities using sensors, such as integrating design files as input to the machines, engineering analysis preproduction, process planning of the supply chain using an additive process or subtractive process, assembly planning, scheduling of resources, simulation, and ongoing process-related data capture. The IoT modules that can be used in advanced manufacturing include data modeling a virtual reality simulation and generating optimal assembly sequences, and camera monitoring components. In a futuristic scenario, interconnected machines of suppliers and customers would interact using M2M communication to request for materials and services as and when minimum threshold or triggering events are generated. Green manufacturing can also occur by bundling products and services together (Bitner, 1997) so that the need to replace products with newer ones is less, and manufacturing companies can move toward service orientation. Customer needs identification is necessary for futuristic manufacturing where lot size of one may be required and this requires customer proximity which may be possible using the IoT sensors embedded in products such as water purifiers or laser printers giving real-time data about usage and other preferences and operating parameters. These can help in new product development, while sensors built in report the health of the equipment and the consumables, which can alert the manufacturer to replenish the consumables. The architecture of such deployments of the IoT consists of hardware (sensors and actuators), middleware, and analytics (Gubbi, 2013). The IoT-based inventory management solutions, including RFID or Bluetooth Low Energy (BLE) beacons, can help manage inventory without human intervention by transmitting data on a real-time basis relating to locations of inventory items, broadcasts on expiry dates, and even assist robotic arms to identify material based on specific SKUs for packing and shipping. Energy management using the IoT sensors can help detect machine-level energy consumption, correlate the changes to machine breakdown trends, and also eliminate wastage by shutting off power when not required based on motion sensors. The IoT-based actuators and motors can also optimize the efficiency level of solar panels by automatically aligning to sun movements. The biggest advantage is of real-time monitoring of energy requirements and consumption rather than a post reporting in the form of power bills. This reduces unbilled energy and also reduces costs proactively. The IoT can help both manufacturers and enforcement agencies by using gaseous detection sensors, water quality detection sensors, and water flow meters in place to detect water content and also gaseous discharge in different locations. Flow meters can help maintain data on total input water and processed water giving real-time statistics on a firm's polluting index. Future industrial systems are linked back and send data to the manufacturer, to work on their service schedule, which traditionally has been time driven (months or years) or usage driven (kilometers or working hours) and not really need driven. They will use the IoT for remote maintenance and increased efficiency to reduce their carbon footprint due to travel and energy saving. The concept of digital twins also saves wastage in the production process by reducing faults in the final product, by replicating the product in a digital form using sensors. Industries can get insights from these sensors' data of the entire working mechanism and the accuracy and effectiveness of each stage. Production systems are prone to failures, consolidation of the IoT with deep-learning methods can make them self-healing as they can take decisions and repair their failures. The IoT devices can also control devices like pumps according to flow of fluid required, and allow remote execution of the production process and batches. Calibrating wastewater treatment generated by the manufacturing process can become more analytical using data captured by the IoT sensors of values of pH value, temperature, flow, and chemical composition and ensure organizations stay within regulatory limits (Becker, 2018). Process costs can be lowered as plants can work on scaled-down volumes to stay within emission control limits, some of which are specified for fixed time periods. Engineering and inspection areas can benefit using the IoT imaging and simulation modeling and innovative models can be developed where product improvement is possible using the IoT sensors with additive manufacturing capabilities to incorporate customer feedback on a real-time basis and move from a lab-centric design approach to real-time data focused and customer-centric design. Rapid prototyping is now being deployed in manufacturing firms to create models using computer-aided designs, thereby enabling production of a limited number of parts for pilot testing. Additive manufacturing is further facilitating this by offering firms the capability to design blueprints in their laboratories and actually manufacture at customer premises. With digitally controlled additive printing, it may be possible in future to replace data inputs using the IoT sensors and devices capturing design-related data elements and transmitting it to labs for continuous product improvements and provisioning in new features.

9.5 Green IoT

Designing for an IoT-enabled network architecture demands an organic approach as compared with traditional IT systems involving networking because of the extreme type of communication requirements and length and breadth of the myriad of devices to be connected. The connections at the end points of the network integrating these devices will be "low fidelity," low-speed, lossy, and intermittent. Also, the communication will be M2M, transmitting small bits of data, unlike the traditional internet that involves transmitting large packets of data but in longer intervals. With billions of the IoT devices about to dot the landscape through connected objects, the power consumption will be huge along with e-waste generation. Hence, green IoT is being promoted worldwide by leading adopters.

9.6 Conclusions

There is pressure on manufacturing organizations to adopt green manufacturing on account of three compelling factors. One is purely economic, where reduction of energy costs and wastages to material can influence the balance sheet positively. Another is the regulations in different countries, which have defined strict emission laws and sewage discharge in addition to noise pollution, all of which are unwanted by products of any manufacturing process. The last is social pressure as organizations compete to meet their social liabilities. The IoT adoption at different stages of manufacturing can help measure and create awareness about productivity, energy efficiency, and process efficiency at a task level and machine level by capturing real-time data and transferring it to integrated systems, which can then provide this as an insight for the next planning or execution cycle. Inventory monitoring using the IoT systems with its capabilities of providing locational information and contextual information can be streamlined and optimized, leading to traceability of time and energy spent in each of the tasks in the manufacturing process. Manufacturing also has by-products and waste products such as emission and discharges in gaseous and liquid formats and each country has regulations in place to control these, and the cost of noncompliance can be significantly high. The IoT sensors as showcased in this chapter can help control the environmental effect of manufacturing by providing real-time data on effluents. Similarly, the use of the IoT in green supplier training and enforcement of practices at the supplier end and monitoring the results have also been explained. The IoT powered collaboration can also bring together virtual suppliers and customers, thereby reducing the transportation environmental impact.

References

Bai, C. A. 2010. Green supplier development: Analytical evaluation using rough set theory. *Journal of Cleaner Production*, 18, 1200–1210.

Banerjee, S. 2002. Corporate environmentalism: The construct and its measurement. *Journal of Business Research*, 55, 177–191.

Becker, A. 2018, January 23. Digitallistmag. *promise-of-iot-go-green-with-greater-efficiency-05790019*. https://www.digitalistmag.com/iot/2018/01/23/promise-of-iot-go-green-with-greater-efficiency-05790019

Bitner, M. 1997. Services marketing: Perspectives on service excellence. *Journal of Retailing*, 73, 3–6.

Bunse, K. V. 2011. Integrating energy efficiency performance in production management: Gap analysis between industrial needs and scientific literature. *Journal of Cleaner Production*, 19, 667–679.

Daily, B., Huang, S.-C. 2001. Achieving sustainability through attention to human resource factors in environmental management. *International Journal of Operations & Production Management*, 21, 1539–1552.

EIA. 2010. *EIA: Annual Energy Review*. EIA.

Gavronski, I. 2012. Resources and capabilities for sustainable operations strategy. *Journal of Operations and Supply Chain Management*, 5, 1–20.

Giovanni, M. A. 2013. Using Internet of Things to improve eco-efficiency in manufacturing: A review on available knowledge and a framework for IoT adoption. In M. T. C. Emmanouilidis (Ed.), *APMS 2012* (pp. 96–102). IFIP International Federation for Information Processing.

Gubbi, J. B. 2013. Internet of Things (IoT): A vision, architectural elements, and future directions. *Future Gen Comp Systems*, 29, 1645–1660.

Gutowski, T. D. 2006. Electrical energy requirements for manufacturing processes. *13th CIRP International Conference on Life Cycle Engineering.*

Haller, S. K. 2009. The Internet of Things in an enterprise context. In J. F. Domingue (Ed.), *FIS 2008. LNCS* (pp. 14–28). Heidelberg: Springer.

Hill, M. K. 2004. *Understanding Environmental Pollution*, 2nd edn. New York.

Krause, D. 1997. Supplier development: Current practices and outcomes. *Journal of Supply Chain Management*, 33, 12–19.

Meyer, G. W. 2011. Production monitoring and control with intelligent products. *International Journal of Production Research*, 49, 1303–1317.

Michaloski, J. S. 2011. Analysis of sustainable manufacturing using simulation for integration of production and building service. *Compressed Air*, 93–101.

Muller, D. M. 2007. An energy management method for the food industry. *Applied Thermal Engineering*, 27, 2677–2686.

Park, C. K. 2009. Energy consumption reduction technology in manufacturing – A selective review of policies, standards, and research. *Precision Engineering and Manufacturing*, 10, 151–173.

Poon, T. C. 2011. A real-time production operations decision support system for solving stochastic production material demand problems. *Expert Systems with Applications*, 38, 4829–4838.

Ramakrishnan, R. L. 2016. Smart electricity distribution in residential areas: Internet of Things (IoT) based advanced metering infrastructure and cloud analytics. *2016 International Conference on Internet of Things and Applications (IOTA)*. Pune, India: IEEE.

Reich-Weiser, C. V. 2010. Appropriate use of green manufacturing frameworks. *Laboratory for Manufacturing and Sustainability.*

Vijayaraghavan, A. D. 2010. Automated energy monitoring of machine tools. *CIRP Annals—Manufacturing Technology*, 59, 21–24.

Weinert, N. C. 2011. Methodology for planning and operating energy efficient production systems. *CIRP Annals—Manufacturing Technology*, 60, 41–44.

Zhang, Y. Q. 2011. Real-time work-in-progress management for smart object-enabled ubiquitous shop-floor environment. *International Journal of Computer Integrated Manufacturing*, 24, 431–445.

10

User Perceptions

10.1 Introduction

As more and more organizations embrace Industrial IoT (IIoT), the potential socioeconomic impact and the readiness in terms of leadership, technology, and capabilities need to be measured. We understand the current state of the Indian organization, and suggest a framework for implementation in industrial setups including systems thinking, business models, and strategic considerations. More than 30 billion objects will be connected under the IoT revolution by 2021 (Ericsson, 2015). The IoT will connect not only things but also people, process, and data (IDC, 2015) with an economic impact of over $1 trillion by 2022. The IoT will influence business operations, society, and daily life, bringing together people (Guo, 2013), technology (Rose, 2015), objects (Atzori, 2010), data (Borgia, 2014), and processes (Shin, 2014). The IoT systems will bring in sensors that can accurately sense the immediate environment. These can measure temperature, help the elderly by way of path finding, help manufacturing companies by integrating assets and making them connected with each other and autonomous, help transportation logistics with geopositioning and location sensing, help energy saving with real-time monitoring of workloads, aid healthcare using ingestibles equipped with sensing elements, and many others. The actors in this IoT world would include people, systems, business, and governments, all speaking to each other (Anastasios, 2016). Organizations worldwide are developing standards for the IoT systems, including W3C (the World Wide Web Consortium), Global Standards Initiative (GSI-IoT), and the Open Interconnect Consortium. The IoT has found practical usage in a number of consumer/governance areas such as healthcare (Islam, 2015), education (Papamitsiou, 2014), home (Stojkoska, 2017), and industrial/business areas from smart building (Pan et al., 2015), manufacturing (Perera, 2014), smart energy (Ramakrishnan & Gaur, 2016), transportation, retailing, and logistics, vehicles, agriculture and livestock, and even in governance domains such as smart cities planning, utilities, and environment monitoring. However, there exists challenges from the technology aspect (accuracy and quality of sensors and devices), society acceptance aspect (topics such as culture and privacy), business issues (including alliances, extended partners, and updating required), and human resistance as objects become more active and autonomous in operations.

10.2 The IoT-UPM (User Perceptions Model)

In this section, we understand the factors responsible for any users in the consumer space (Consumer IoT) to use the IoT-related technology. We classify the factors into four broad areas (Economides, 2008).

- System-related attributes that deal with the application including hardware, software, and people aspects, including user friendliness and efficiency.
- User-related characteristics include the IoT literacy levels, adeptness in using computers, or related devices and preferences.
- Service-provider-related features including brand name, industry, size of operations, and the level of customer support.
- Overall environment including other ecosystems, the IoT systems, integration extent, etc.

These factors interact between themselves and with each other and also influence the willingness of the consumer to adopt this technology or product and to continuously use it. Some of these factors and their extent of interaction can be measured, while some can just be objectively stated. The ones that can be measured are real context while the others are perceived context (Figure 10.1).

The model derives strength from the traditional models such as the Technology Acceptance Model (Davis, 1989) or the ISO Quality models (ISO/IEC 24765, 2008). Table 10.1 defines the attributes that determine the user and the IoT system interaction and the user perceptions.

Table 10.2 describes the attributes specific to the operations of the IoT system.

Table 10.3 shows the attributes that define the success of results derived from the IoT systems.

The 31 identified factors listed in Tables 10.1 through 10.3 influence the user's perception and attitude and ultimately their acceptance of an IoT system and can be considered to constitute the building blocks of a user perception model. This model was put to test in India over a survey of 100 participants who were health conscious and were asked to evaluate the use of a smart wearable for real-time healthcare tracking. The wearable industry has been plagued by concerns on security of sensitive data such as healthcare and identity tracking of the individual and their lifestyle and the fear of this data getting used for detrimental reasons including denial of health insurance policies (Flaherty, 2014). Usage, however, is also increasing among the younger age group

FIGURE 10.1
Factors affecting the IoT systems usability.

TABLE 10.1

The IoT System Interaction Related to User Perception

User Friendliness	This refers to the ease of using the IoT system, including the intuitive appeal, minimized number of steps, ease of carrying or installing or updating the device, and the effort required to operate the system.
Accessibility	This refers to the degree to which the system can be used by different users with varying capabilities and diverse characteristics can easily use the system.
Awareness	Visibility about the system's functioning and ability to observe, accurately and timely, usage and billing of the proposed system.
Appeal	The system's user experience in terms of visualization and auditory capabilities.
Support	Support received by a user of an IoT system in terms of help desk, online guidelines and pointers, and automated support for regular operations.
Ubiquity	The uptime and continuous running capabilities without failures, thereby ensuring a ubiquitous and seamless experience.
Personalization	Ability to customize the device and its services including interfaces, operations, and functions as per customer's individual preferences.
Control	Ability to control during the course of usage, resources, and data and do corrections if required.

TABLE 10.2

The IoT Operations and User Perception

Efficiency	Refers to the extent to which an IoT-enabled system can help an individual with the level of performance and resource utilization.
Functionality	Ability of the system to provide necessary functions for execution of the target business process objectives.
Performance	User experience of speed and throughput benchmarked against user perceptions.
Reliability	Ability of the system to handle exception cases, errors, and malfunctions so that user operations is not interrupted.
Availability	Ability to continue with operations using the IoT system in case of external factors such as electricity or network disconnection.
Energy	This relates to energy efficiency in operating an IoT system and energy saving modes applicable.
Security	The perception of the user from the security aspect including unauthorized access risks and losses.
Safety	Perception in safety issues including risk of losses or injuries or negative outcomes from the system usage.
Endurance	User perception on how long this technology will survive obsolescence and can be used.
Flexibility	Degree of modifications possible in the IoT system, for achieving tailor-made objectives.
Compliance	Ability of the IoT system to follow standards, laws, and regulations.
Interoperability	The extent of coexistence and integration with other systems and processes.
Integration	Extent of sharing a common or shared infrastructure and resources with other systems.
Autonomy	Ability to perform actions independently, and communicate results or take actions based on specific scenarios.
Replicability	Ability to substitute the system with alternate competing systems and avoiding lock in to a specific product or service.
Extensibility	Ability to upgrade facilities on the fly using custom software and hardware to address new or additional requirements.
Scalability	Ability to meet extra demands, by increasing capabilities and adding performance to produce more outcomes.

leading to some societal concerns (Hamblen, 2015). Information privacy and data security concerns could affect the individual behavior toward adopting technology (Stewart & Segars, 2002). The above 31 factors in this survey were found to be ranked significantly higher than other set of 20 variables introduced to render validity to the survey variables and significant association was found between them. The top two factors that define the user's perception of success for the adoption of healthcare wearables were usefulness and effectiveness and all participants felt that the device should be useful in meeting the cause of healthcare monitoring and should be effective in terms of measurements, which could be similar with very less margin of error to clinical or professional instruments. The user interaction with the wearables were largely influenced

TABLE 10.3

The IoT System Success Factors

Usefulness	The extent to which results are achieved against stated objectives, including solving a problem or producing a desired state or result.
Satisfaction	The extent the user feels to which using the IoT system would be able to meet the user requirements completely and on time.
Effectiveness	The extent to which the solution is effective and meets required objectives on the field run with accurate results.
Value	The extent to which usage of the system adds value to a user operation such as reducing time or ensuring better quality. Overall process and product/service value addition derived due to adoption of the system.
Privacy	The ability and the amount of control a user has over information concerning their financial, health, age, or demographics and other disclosures.
Collaboration	The ability of a user and the system to create a platform where people can share information, communicate, and connect.

by the user friendliness of operations and the ubiquitous nature, that is, availability anywhere and always. Sales were influenced by appeal of the device and its user esthetics. This is in-line with the observations of similar research as evidenced by the literature review (Yang et al., 2016) and the issues related to privacy and security concerns are also identified in the survey (Malhotra, 2004).

10.3 The IoT Technology Acceptance Model

In this section, we enumerate the factors responsible for business organizations and their user departments to adopt the IoT technology in their processes and works. The Technology Acceptance Model (TAM) is well established to gauge the efficiency of IT-related technologies in a workplace. Systems are still unutilized even though hardware and software advancements have taken place exponentially and the same is true for the IoT as well and since investments in the IoT system can be huge, the effect of low returns will be highly counterproductive. Significant empirical and theoretical support exist for the TAM, which highlights that the usage of IT systems is dependent on perceptions of the user toward usefulness of the system and the ease of use (Davis, 1989). This was further updated to the TAM2 model, where five factors were added to the original model: subjective norm, image, job relevance, output quality, and result demonstrability (Venkatesh & Davis, 2000). There have been alternate literature also including Theory of Reasoned Action and Theory of Planned Behavior, but they have been less successful (Venkatesh & Davis, 1996). As per TAM, effects of external variables are mitigated by the above two. However, in the case of the IoT where human intervention is limited and objects are autonomous and self-communicating, these two

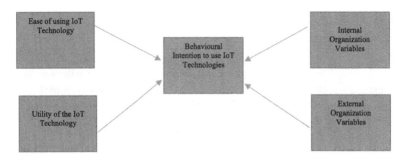

FIGURE 10.2
Conceptual IoT-TAM model.

dimensions are diluted. The following model is proposed, which is referred to as the IoT-TAM to adapt the above model in line with the IoT technology and projects. It retains two of the older variables namely the perceived usefulness of the IoT and the perceived ease of use of the IoT while introducing two more variables, the internal and external factors. This is because the IoT project implementations are very complex and not strictly technological but involve process, people, data, and integration among other things (Figure 10.2).

The end-user perception, that using the IoT technology is easy, can be defined by how useful they find it for their day-to-day operations and in meeting the business requirements in a user-friendly manner, which translates to the amount of training required, the chances of a user making a mistake that is not controlled or shown as an alert by the system leading to loss of data work productivity, and the availability and skill requirements. The second aspect is the utility of the IoT technology in performing operations that may be measured by productivity increase, better controls, and increase in efficiency and throughput, better operations or maintenance, and data capture for further analysis. External organization variables include those that the surrounding ecosystem brings in as an influencing factor such as requirements from customers and suppliers to integrate digitally or have real-time visibility or collaboration. It is also affected by similar firms adopting and achieving results through IoT technology adoption. Measures of internal organization variables include maturity and extent of modernization of traditional IT systems, ability of assets to be digitally connected, and products to have a digital footprint in addition to the physical footprint, financials budgets, and the role of decision-makers from business and IT, which is specific to the organization. These factors are also interlinked and a strong set of internal variables in the organization can significantly improve perception relating to the ease of use, while a strong set of external variables provide economically significant and business-friendly use cases either from external parties or from related parties, thereby improving perception about the utility of the IoT technology. However, while internal variables are very much focused at the firm level, external variables will tend to cluster around a specific industrial domain or size of the firm or the nature of activities.

To validate this model, a field survey was conducted on a population of business and technology decision-makers in India, with 171 participants (Ramakrishnan et al., 2017) representing different organizations with diversified business domains and of different sizes. The results showed that the assumption of the conceptual model above could be validated through the field trials. Firms that had a robust internal IT and a progressive technology driven management were able to go for the IoT adoption much sooner than others and could also derive more value from their IoT adoption. Similarly, industries such as telecom and utilities, which are technology intensive, and consumer white goods above a specific value, which can make integration of the IoT devices economically feasible without impacting profitability, have significantly derived more benefits from the IoT adoption. Similarly, manufacturers with assets that can be digitally enabled to combine operational technology with digital technology have a higher success rate while adopting IoT and can find better use cases. Firms that have good traditional IT systems also have better trained and knowledgeable end users who are able to understand and follow the practices required for the IoT project adoption and implementation.

10.4 The IoT Maturity Model

Based on a profiling of numerous business organizations done by the authors, this section presents the IoT maturity model that can be used by organizations to benchmark themselves while also formulating the next steps in their journey toward adopting the IoT. Currently, there are two well-recognized models for Industry 4.0 maturity: the IMPULS and the PwC model.

The IMPULS model has the following six dimensions and attributes:

1. Strategy and Organization, which includes focus on Strategy, Investments, and Innovation Management
2. Smart Factory, which includes Digital Modeling
3. Smart Operations
4. Smart Products
5. Data Driven Access
6. Employees

The PwC model constitutes four layers: Digital Novice, Vertical Integrator, Horizontal Collaboration, and Digital Champions. Each depict a progressively increasing layer or Industry 4.0 maturity based on predefined attributes.

The attributes are as follows:

1. Business Models adopted and Product and Service portfolio offered
2. Market and Customer Channels maturity and the data being gathered to understand their interaction
3. Integration of process and sections such as to what extent manufacturing operations in a company is integrated with the supply chain
4. IT and Digital architectures
5. Compliances and risk management
6. Organizations ability to adopt digital in terms of its workforce and internal culture.

As was evident in Section 10.3, the IoT-TAM model lists internal factors specific to a firm as being instrumental to the IoT adoption and improved perception on ease of use. This section identifies a conceptual model using a specific set of internal factors.

10.4.1 Traditional IT Systems

IT systems influence and are a key requirement for the IoT enabling an organization (Fan et al., 2013). Organizations having software systems like the management execution systems, ERP and CRM systems, and business intelligence and asset management software based on RFID or barcode will have a stronger case of adopting the IoT. Here again, organizations can be divided into five progressive tiers or strata. The first layer being minimal IT such as desktop standalone systems, word processing, and spreadsheet programs. The last layer being of highly advanced analytical systems with robotic controls and a fair degree of autonomous data-driven operations. Similarly, the hardware infrastructure varies from standalone desktops to high-end engineered systems while networks can vary from simple local area networks (LAN) or standalone systems to highly complex virtual LAN and next-generation digital networks, such as Quality of Services. The next aspect would be the internal strength of IT resources (both employees and contractual) and their capabilities to implement IT solutions and the strength of the business to model the use case in IT terms.

10.4.2 People

People and devices need to coexist in a mutually symbiotic manner in an IoT-enabled firm (Turunen et al., 2015), with human computer interfaces playing an important role. The IMPULS and PwC models for Industry 4.0 also highlights this aspect of people and culture. A better skilled force that understands and appreciates the use of traditional data and IT systems will be in a better position to adopt IoT systems, which also require autonomy to be delegated to machines and role sharing.

10.4.3 Connected Assets

Connect Assets refers to the manufacturing setup of the organization in terms of plant and machinery, converting assets, utilities assets like chillers and boilers, heating ventilation, and cooling systems. Traditionally, many of the manufacturing assets were focused on production output and had limited digital connectivity and operated as standalone machines with limited processing power and memory footprint (Maasouman & Demirli, 2015). However, for optimal IT-OT connectivity (information technology and operations technology), and to enable data transmission of machine-to-machine (M2M) communication, assets must be capable of data broadcast transmission about their operating state and surroundings leveraging on digital networks and this data could be stored in traditional IT systems and analytics for processing machine data for insights.

10.4.4 Digital Products

Traditionally, products had a physical aspect, which was the primary differentiator. Of late, products are acquiring a digital aspect, which is actively contributing the revenue. This can be provisioned on the fly, such as the usable capacity of the engine in a Tesla car.

10.4.5 Financial Feasibility

For a technology project, RoI, cost of deployment, and profitability are the driving factors (Chovichien & Nguyen, 2013). For any IT project, including the IoT, feasibility needs to be established from inception to closure. Common ways to achieve this include using modeling or setting indicators (Stevens, 1993). There have been studies to determine profitability of the IoT projects for specific companies such as logistics (Henning, 2009) or for understanding the economic benefits of having adopted the IoT (Bottani & Rizzi, 2008). The financial feasibility depends on market opportunity and visible demand, potential revenue generation, strategic alignment to goals, time to market, and strategic importance for the firm (Jovanov et al., 2005) (Figure 10.3).

10.5 Insights Based on a Field Survey

Based on a field survey assessment from a top IT and business decision-maker, the following insights emerge relating to the IoT technology wave in the coming few years:

FIGURE 10.3
The IoT maturity indicators of a firm.

- The opportunities for IoT seem abundant and largely unexplored, however lack of governance standards and interoperability between machines manufacturers along with the fact that most of the machine assets still date to the last century (pre-2000) is creating a bottleneck for adopting the IoT. Assets are still not digitally enabled for connectivity and in the absence of proven business models specific to an industry, there is reluctance for going for an expensive overhaul.

- Scalability challenges in terms of data volumes and connectivity requirements need an architectural shift and realignment and hence endpoint technologies have to evolve disruptively. The current Wi-Fi or LAN model of connecting desktops and mobile computers will be ineffective in terms of compute power, resource requirement, and energy consumption. This will require consolidation of the so many fragmented protocols and standards.

- Business outcomes will also require supporting technologies like machine learning and artificial intelligence (AI) to work on the IoT data and give insights for actionable outcomes.

- There is a gradual transition from the initial exploratory phase (or the curiosity phase) of the IoT to an execution phase as people have started taking the first steps toward their IoT project and for assessment of their current IoT maturity as an organization.

- The IoT technology can be divided into five sections: relating to sensor and end points, relating to communications and network, relating to

FIGURE 10.4
Broad areas of the IoT.

security and privacy, relating to data intelligence and analytics and, machine learning and AI (Figure 10.4).

- The apprehension of success in the IoT projects is dismally low in the minds of those surveyed, due to the perception that the technology stack has not matured, and there are inherent deficiencies as of now—security, privacy, risks, and liabilities being the major ones.

- In terms of sensors, there has been some research (Gartner, 2017) on innovative developments to make the IoT more accessible, focusing on desktop level development tools for prototyping the IoT solutions to model new business streams with minimal technology investments, support for the IoT cost optimizations strategy based on a checklist and impact assessment, software solutions for multimodal approach to smart cities, and tools for wearable strategies.

- There is still doubts on how to monetize data without scaring away the users who generate data, while some compelling cases exist for contextual data specific to locations and time but the fact that an individual can be traced back using this data and their behavior predicted is emerging as one of the biggest stoppers.

- In manufacturing, the concept of digital twins is evoking interest; however, the effectiveness of complex machine modeling is still limited as software maturity to adopt to years of manufacturing practices are still developing and there are still margin of errors.

- Energy management using the IoT devices is being adopted to address the issues of costs and environmental impact minimization.

10.6 Conclusion

This chapter has presented three models relating to IoT. The first is related to user perceptions model, and the factors are identified and defined that influence it, which can be used by a solution provider to make their offerings more lucrative and acceptable in higher probability to end users. The second model relates to IoT-TAM for enterprise adoption of the IoT and perception, which identifies the set of four variable groups as an extension to the TAM model and includes internal factors and external factors that are specific to a firm and the specific industrial domain, respectively. Subsequently, a five-stage maturity model has been defined for the maturity level assessment of a firm and the degree of adoption possibility and actual adoption of these factors with respect to the IoT technology. In the end, we summarize with an insight report as gathered by professional literature for the IoT technology in the near future.

References

Anastasios, A. E. 2016. User perceptions of internet of things (IoT) Systems. *13th International Conference on E-Business and Telecommunications*, Springer: Lisbon, Portugal, July 26-28, pp. 3–20.

Atzori, L. I. 2010. The internet of things: A survey. *Computer Networks*, 54, 2787–2805.

Borgia, E. 2014. The internet of things vision: Key features, applications and open issues. *Computer Communications*, 54, 1–31. http://dx.doi.org/10.1016/j.comcom.2014.09.008.

Bottani, E., Rizzi, A. 2008. Economical assessment of the impact of RFID technology and EPC system on the fast-moving consumer goods supply chain. *International Journal of Production Economics*, 112(2), 548–569.

Davis, F. D. 1989. Perceived usefulness, perceived ease of use, and user acceptance of information technology. *MIS Quarterly*, 13(3), 319–340, doi: 10.2307/24900.

Maasouman, M. A., Demirli, K. 2015. Assessment of lean maturity level in manufacturing cells. *IFAC-PapersOnLine*, 28(3), 1876–1881.

Economides, A. 2008. Context-aware mobile learning. In Lytras, M. C. (ed.) *WSKS CCIS*, vol. 19, (pp. 213–220). Heidelberg: Springer., doi: 10.1007/978-3-540-87783-7_27.

Jovanov, E., Milenkovic, A., Otto, C., de Groen, P. C. 2005. A wireless body area network of intelligent motion sensors for computer assisted physical rehabilitation. *Journal of NeuroEngineering and Rehabilitation*, 2, 6, doi: 10.1186/1743-0003-2-6.

Ericsson. 2015. *Ericsson Mobility Report*. Retrieved August 12, 2018 from http://www.ericsson.com/res/docs/2015/mobility-report/ericsson-mobility-report-nov-2015.pdf

Fan, P. F., Wang, L. L., Zhang, S. Y., Lin, T. T. 2013. The research on the internet of things industry chain for barriers and solutions. *Applied Mechanics and Materials*, 441, 1030–1035. https://doi.org/10.4028/www.scientific.net/AMM.441.1030.

Flaherty, J. L. 2014. Digital diagnosis: Privacy and the regulation of mobile phone health applications. *American Journal of Law and Medicine*, 40(4), 416–441.

Gartner. 2017. *Cool Vendors for 2017: The Digital Nail Gets Hammered, So Be the Hammer.* Retrieved September 1, 2018 from Gartner Webinars: https://www.gartner.com/webinar/3710220?ref=solrSearch&srcId=1-5418502758.

Guo, B. Z. 2013. Opportunistic IoT: Exploring the harmonious interaction between human and the internet of things. *Journal of Network and Computer*, 36(6), 1531–1539.

Hamblen, M. 2015. As smartwatches gain traction, personal data privacy worries mount: Companies could use wearables to track employees' fitness, or even their whereabouts. *Computerworld*, https://www.computerworld.com/article/2925311/as-smartwatches-gain-traction-personal-data-privacy-worries-mount.html.

Henning, B. G. 2009. Evaluation of RFID applications for logistics: A framework for identifying, forecasting and assessing benefits. *European Journal of Information Systems*, 18(6), 578–591.

IDC. 2015. *Worldwide and regional Internet of Things (IoT) 2014–2020 forecast: A virtuous circle of proven value and demand*. Retrieved June 15, 2018 from http://www.idc.com/downloads/idc_market_in_a_ minute_iot_infographic.pdf

Islam, S. K.-S. 2015. The Internet of Things for health care: A comprehensive survey. *IEEE Access*, 3, 678–708. doi:10.1109/ACCESS.2015.2437951

ISO/IEC 24765. 2008. Systems and Software Engineering Vocabulary.

Malhotra, N. S. 2004. Internet User' Information Privacy Concerns (IUIPC): The construct, the scale, and a causal model. *Information Systems Research*, 15, 336–355.

Chovichien, V., Nguyen, T. A. 2013. List of indicators and criteria for evaluating construction project success and their weight assignment, 4th International *Conference on Engineering Project and Production*.

Pan, J., Jain, R., Paul, S., Vu, T., Saifullah, A., Sha, M. 2015. An internet of things framework for smart energy in buildings: Designs, prototype, and experiments. *IEEE Internet of Things Journal*, 2(6), 527–537.

Papamitsiou, Z. E. 2014. Temporal learning analytics for adaptive assessment. *Journal of Learning Analytics*, 1, 165–168.

Perera, C. L. 2014. A survey on Internet of Things from industrial market perspective. *IEEE Access*, 2, 1660–1679.

Ramakrishnan, R., Gaur, L. 2016. Feasibility and efficacy of BLE beacon IoT devices in inventory management at the shop floor. *International Journal of Electrical and Computer Engineering*, 6, 2362–2368.

Ramakrishnan, R., Singh, G., Gaur, L. 2017. Internet of Things – Technology Adoption Model in India. *Journal of Science and Technology Pertanika*, 25(3), 835-846.

Rose, K. E. 2015. *The Internet of Things – An overview – Understanding the issues and challenges of a more connected world*. ISOC. Retrieved June 15, 2018 from http://www.internetsociety.org

Shin, D. 2014. A socio-technical framework for Internet-of-Things design: A human-centered design for the Internet of Things. *Telematics and Informatics*, 31, 519–531, http://dx.doi.org/10.1016/j.tele.2014.02.003.

Stevens, F. L. 1993. *User friendly handbook for project management: Science, Mathematics, Engineering and Technology Education*. Washington DC: National Science Foundation.

Stewart, K., Segars, A. 2002. An empirical examination of the concern for information privacy instrument. *Information Systems Research*, 13(1), 36–49.

Stojkoska, B. T. 2017. A review of Internet of Things for smart home: Challenges and solutions. *Journal of Cleaner Production*, 140, 1454–1464.

Turunen, M., Sonntag, D., Engelbrecht, K. -P., Olsson, T., Schnelle-Walka, D., Lucero, A. 2015. Interaction and Humans in Internet of Things. In: Abascal, J., Barbosa, S., Fetter, M., Gross, T., Palanque, P., Winckler, M. (eds) Human-Computer Interaction – Interact, volume 9299. Springer, Cham: Switzerland, doi: 10.1007/978-3-319-22723-8_80.

Venkatesh, V., Davis, F. D. 2000. A Theoretical Extension of the Technology Acceptance Model: Four Longitudinal Field Studies. *Management Science*, 46(2), 186–204.

Venkatesh, V., Davis, F. D. 1996. A model of the antecedents of perceived ease of use: Development and test. *Decision Sciences Summer*, 27(3), 451–481.

Yang, H., Yu, J., Zo, H., Choi, M. 2016. User acceptance of wearable devices: An extended perspective of perceived value. *Telematics and Informatics*, 33, 256–269. doi: 10.1016/j.tele.2015.08.007.

11

The Future of the Internet of Things

11.1 Introduction

The Internet of Things (IoT) is connecting devices with diverse and varied uses like meters, lights, consumer white goods, electronic items, media devices and set top boxes, automobiles, and many machines of industrial usage. The IoT basis is on the evolution and existence of ubiquitous computing, with software and hardware present everywhere (Weiser, 1991). While domestic appliances will focus in miniaturization, low-power consumption, and security algorithms inside devices, industrial applications of IoT will be more focused on integration of non-TCP and TCP networks with gateways for real-time data transfer. The future of manufacturing will require the ability to respond to dynamic customer demands, become competitive without having physically owned manufacturing facilities, and, hence, operating in virtual production environments. The future of manufacturing will be driven by a number of design considerations such as the ability to respond to dynamically changing demands of the customer and business, handling a higher degree of customer demands, and adapting to rapidly evolving production models from current mass production to lot size of one production, which is highly tailor-made from currently making stock to an engineer-to-order option. However, these requirements should not lead to spiraling of manufacturing costs or take a higher time to service than the current options so that overall profitability can be retained. Current owners of facilities may offer only machines on a service or pay per use model while people with design and ideas may just get to use others' infrastructure while fulfilling orders. In other words, the world of manufacturing would become loosely coupled. Evolution of new business models for manufacturers with higher pricing margins may be possible as they shift to tailor-made products. Adoption of outsourced production in leased facilities which support usage-based billing will enable manufacturers to real-time collaborate with different facilities providers, and drive innovation through an optimal mix of machinery, something which is currently not possible in today's world. Today, the major limitation that a manufacturer faces is firstly the capital-intensive nature of machines and, secondly, the limitation of a particular manufacturing asset in terms of capabilities, production capacities, and product mix possible. All this might become a thing of the past as manufacturers work on OPEX models sourcing material from nearest proximity and then go for hybrid production models. Future factories will be machine-man driven, a higher emphasis for machines as they become smarter and more autonomous and totally data driven. The current role of humans in operating machines, such as forklifts or cranes, will be eventually replaced by robots with their wider capabilities and increasing workhours. Already, we have numerous cases of huge warehouses totally manned by robots or assembly lines with fully autonomous operations.

11.2 Hybrid Technologies

The IoT has three main categories of users: (1) domestic, (2) industrial, and (3) public sector services (Sivakumar, 2017). With the upcoming technologies like the IoT—both consumer and industrial—artificial intelligence, robotics, and evolving machine learning provides an open environment where potential business partners and their machines can collaborate and exchange information over a digitally connected network using cyber-physical systems. The ability of new manufacturing assets at the shop floor to integrate operational technology data running on SCADA and PLC systems with the TCPIP network has opened up many avenues for mining this data. This can be used for production planning, remotely controlling assets, predicting failure rates, and completing preventive maintenance. This also enables business models where remote machines can be operated by end customers such as scheduling batch runs or dynamically changing production runs in between based on production priorities and idle time availability. This way, visibility can extend right across the value chain, something not possible today, so right from the initial provider of the raw material to the final customer, each would be able to see and report defects. This becomes particularly interesting in a scenario where an automobile manufacturer ("aggregator or discrete manufacturer") buys a gear box or tires from a vendor ("supplier") who in turn sources rubber from another supplier ("initial raw material supplier") and keeps the automobile with a distributor who further sells to the end customer. Today, when a complaint is raised by the end customer of a defective tire, which wears off after few days of usage, the same can never be traced back to the initial raw material supplier who has no visibility or feedback on their product and hence can bring in no process improvements. With a world of connected assets across the value chain and integration across the manufacturing ecosystem, traceability can be established end to end, leading to process improvements across all entities. The IoT of today will extend to cover process, people, products, and other infrastructure to become Internet of Everything (IoE), thereby streamlining the manufacturing process. The availability of the network spectrum is another area of concern as there are billions of devices that will need a new radio frequency spectrum to communicate since the existing spectrum seems inadequate for mobile devices, but this is far from true (McHenry, 2006) and the spectrum is highly underutilized for space and time. This is planned to be addressed using dynamic spectrum access mechanisms (FCC, 2002), which can play around with the variables space, time, transmission power, etc. This will be implemented using cognitive radio sets (CR), which will connect and dynamically seek spectrum allocation; its current usage is, however, limited due to higher costs and energy consumption, both of which will supposedly improve in the future technology evolution.

11.3 Futuristic Manufacturing Value Chains

Virtual manufacturing applications connecting suppliers and customers, manufacturers, and retailers will define the future, each riding on transparent flow of information. The IoT-enabled factory automation is another area where production facilities equipped with sensors and transmitting real-time solutions enable the other stations (in case of process manufacturing) or assembly lines (discrete) to be better enabled to handle incoming load. Similarly, a breakdown situation will be communicated by the machinery so that material movements can take place to a different destination, avoiding slack or slippages in the production chain. Use of additive manufacturing or 3D printing is yet another area where instead or routine logistics, material will be designed centrally but printed remotely on the nearest additive printing station to the customer, using digitally and secured encryption. Predictive maintenance and ability to forecast breakdowns will be another futuristic area driven by machine data and deep-learning algorithms that would be able to predict failures based on operating parameter values of a specific machine. Another area could be autonomous recycling based on quality parameters wherever recycling is possible, thereby reducing the current tendency of humans to push substandard quality material for customer acceptance. This also means that quality checking of material will become automated in the future (Figure 11.1).

There are substantial risks also associated with such large-scale developments and deployments, primary among them being:

- Limited pilot deployments and test beds with proven benefits, thereby making it more hype-oriented than realistic.
- Failures of earlier architectures like SOA in demonstrating substantial benefits and reducing complexity has resulted in reluctance on part of original equipment manufacturers.

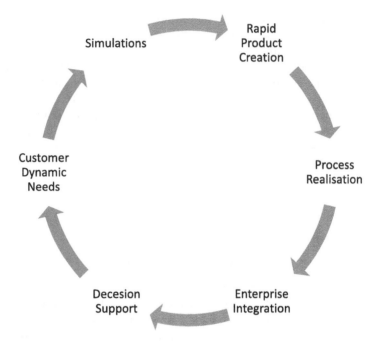

FIGURE 11.1
Virtual factory requirements.

- Long gestation period to enforce the transition is another issue as many assets are legacy and organizations may not have the willingness or capability to change or upgrade without a proven use case.
- Technology maturity is still a gray area as many of the IoT devices fail under extreme industrial conditions. Then there are issues of power consumption. Security is a neglected area in the IoT and migration for legacy protocols are unsupported and nonexistent.
- Investment in R&D in terms of better devices is still in a nascent stage.

Future manufacturing will use technology to shift from mass production to mass customizations, from benefitting from economies of scale to setting up localized production units leveraging outsourced facilities. Focus will shift from cost of unit production to return on capital method, which will be based on data generated from a robust and reliable OT-IT fusion. The trends are already in Europe, which has been driven by Industry 4.0 as key enabling technologies are graduating from laboratories to a commercial model and are driving innovation in the fields of nano and micro electronics, industrial biotechnology, and AMI or advanced manufacturing technologies. Laser-driven precision manufacturing, high performance computing (HPC), and

CPS or cyber-physical systems are now being innovated in small labs and innovation hubs and then economical and feasibility considerations are driving commercial adoption. There is a definite impact on the environment due to the large-scale introduction of the IoT devices; CO_2 emission will increase due to proliferation of these devices from 800 to 1,200 million tons by 2020 (Erricson, 2013). Future IoT devices will be self-sensing with built-in controls for environmental safe guards with the ability to start or shut/hibernate depending on the energy situation as observed in space stations and space navigating robots of today. Future IoT devices will also be more capable of autonomous transmission of data and with limited intelligence built in, which can evolve with time to take decisions.

11.4 Futuristic Consumer IoT

The biggest development will be of connected autonomous vehicles, or self-driven vehicles that can also influence many related industries like insurance, using inbuilt sensors that can provide information about the driving habits and the wear and tear of a vehicle, and thus benefit both the customer and the insurer. Connected vehicles can use either cellular connectivity (car-to-network), a telematics service provider (car-to-TSP), connect to the cloud using Wi-Fi (car-to-cloud), or just be a short-range car-to-car connectivity, which can engage with traffic lights or speed monitors.

The IoT in education along with technologies like AI and cognitive-contextual information will usher in personalized content dissemination to students (Hanan, 2017). Hence, connecting devices will allow tailor-made person-specific monitoring of performance and, thereby, adjusting the speed on teaching dynamically rather than the prevalent class-specific teaching of today. Smart classes and universities will bring a new paradigm of efficiency not observed so far (Lopez Research, 2013).

The IoT in healthcare will be focused on implants and devices that can be consumed so that they can navigate inside our circulatory systems, dispensing medicines to the exact spot of the diseases internally, and then be easily degradable inside our body. Today's medicine, which is more focused on spot assessments (such as blood pressure readings), will become more continuous as wearables become part of our lives. Predictive analysis based on these readings can be used for preventive medications and also for improving quality of life in disabled or elderly patients or those needing critical healthcare or continuous monitoring.

Another area where the IoT holds promise is new product development and innovation. With the IoT sensors on devices generating vast amount of data, it is now possible to measure and gauge attributes that were earlier difficult to measure or could be measured only randomly over a period of time using traditional methods like surveys. These inputs will help improve enhancements to existing product design and functionalities and also help with new product development (Ramakrishnan, 2017). Similar studies have shown the impact of the IoT in supply chain management competencies and product planning. With huge distribution networks, future demand will be for connected objects to rightly forecast demand based on current consumption data and then aggregate it in a hierarchical level in much the same format as it is being done today, where retailers send consolidated demand to distributors who liaison with the original manufacturers through a multilayer distribution network. This will help in having an information-based supply chain demand generation and supply management, with tighter coordination (Harper, 2010).

The future will see the evolution of ecosystem models for the IoT in different industrial domains (Marcel, 2017). Business ecosystems start with the premise that innovations are vital to a firm but needs multiple firms to collaborate, thereby resulting on collaborative networks (Moore, 1993) and requiring collective action (Olson, 1965).

11.5 Emerging Trends

The emerging trends in the IoT development and evolution will be centered across these areas.

11.5.1 Operating Platforms

As the IoT objects start integrating more and more complex sensors and become integrated with standalone capabilities, including hardware, software, connectivity, and intelligence, they will start behaving like platforms, either serving the function of low-level device control and communication operations, or the IoT data acquisition and transformation or the IoT application development.

11.5.2 Standards

New ecosystems will emerge, to facilitate the IoT collaboration. This has to support multiple standards, which may require product upgrades and ensuring the applications support billions of devices as the IoT technology matures. Similarly, industrial IoT will see interoperability come into play as devices from multiple vendors will need to adopt a common set of open standards and not rely on proprietary data formats of standards, which are vendor specific.

11.5.3 Distributed Stream Computing

The IoT devices due to their huge numbers will generate high volumes of data that will need to be transmitted regularly requiring huge bandwidth. Future technology evolution will be able to scale up using distributed stream computing to millions of records that use parallel architectures for real-time analytics on these data volumes.

11.5.4 Operating Systems

Future operating systems for the IoT devices will need to work on low-form factor devices with limited computing power and yet be able to support complex security algorithms like encryption and ensure data privacy and also be able to perform fringe operations at their level. Such real-time operating systems capable of running on small memory footprints will definitely evolve in the future.

11.5.5 Hardware

Current technology hardware like the Intel Edison and Galileo, the Raspberry PI, or the Arduino Uno require shields to connect to hardware equipment. Future hardware will evolve plug and play adaptations, which cannot only add peripheral hardware but also have remote sensing to enable hardware to connect to each other dynamically on demand. Also, hardware needs to have limited power consumption, which is one area where we expect lot of improvements to ensure the IoT technology remains sustainable and capable of mass adoption at low affordable costs.

11.5.6 Networks

Low power networks that can support non-TCP protocols and help in peer-to-peer networks like connected cars or connected industrial machines is another area that needs to see development. With the IoT in machines, latency becomes a very important factor as traditional cellular networks are not designed to meet these challenges, and, hence, networks with wide area coverage and high connection density and low operating costs will evolve with time. The initial low-power WAN were based on proprietary technologies, but recent developments like NB-IoT (narrowband) or LTE-IoT will evolve.

11.5.7 Analytics

A new approach for analytics will be required as huge volumes of data keep piling up, to transform them into actionable insights from raw data. Current analytics based on applying statistical calculations on data as a whole will be replaced with machine learning, which learns up to a certain dataset and then evolves further on the incremental data only. Artificial intelligence models can be improved with large datasets, powered by growth in crowdsourcing and open source analytics.

11.5.8 Wider Scope

The Internet of Things (primarily machine-to-machine communication) has already evolved to the Internet of Everything (machines, people, data, and process) expanding its scope (Cisco). In future, people will be able to connect to the internet in many ways, and people will become nodes on the internet. Also, rather than raw data being transferred as of now, in the future, information or processed data will be transferred based on concepts like fringe computing leading to higher controls. The IoE in environmental monitoring, be it wildlife, pollution levels, noise levels, hunger, and drinking water, will literally map Earth's heartbeat. The Internet of Robotic things, (IoRT) will focus on the many ways that robots can use the IoT technologies to provide advanced capabilities.

11.5.9 Security

The IoT requires IT security to be extended to operational technology (OT), while IT security focuses on application layer, network layer, and data layer and employs methods for authentication, authorization, and provisioning. The OT security will focus on extending these to the shop floor machine level. Traditional threats like DOS will become more profound in the case of the IoT as there will be numerous entry points to access data networks. The blockchain, which is a tamper proof, auditable, and immutable timestamp-based ledger (Kosba, 2016), approach requires significant computational overheads and latency that may not be suitable for the IoT applications that will require real-time information relay. However, with the IoT data privacy and security becoming prominent, the future applications will see the integration of blockchain, which provides decentralized security (Ali, 2017). To minimize the drawbacks of adopting this technology, a lightweight blockchain approach will be needed, which can avoid underlying bitcoins and optimize the underlying algorithms, that is, Proof of Work (POW) or Proof of Stake (POS). POW is compute power intensive while POS is both compute and memory intensive since appending a block in the chain requires solving a complex puzzle. Conventional security algorithms are ineffective in the IoT due to resource consumption, centralization, and privacy issues. Also, with each node broadcasting availability and requiring verification from participating nodes, big scalability issues are created with billions of objects or nodes in the IoT scenario.

11.5.10 Identity Management

The issue of identity management in the future IoT scenario is a very pressing one as users would want their identities to remain secure from others while sharing information as required to solicit replies from other

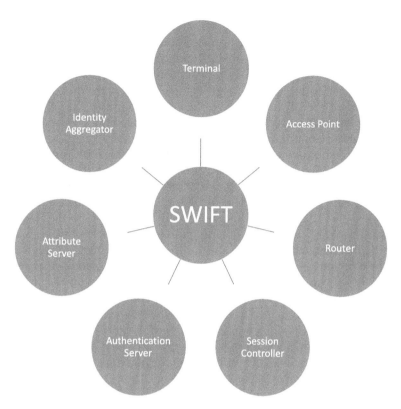

FIGURE 11.2
SWIFT architecture nodes.

business partners. There have been numerous projects for addressing identity management issues such as DAIDOLOS and SWIFT, and some have focused on generating virtual ID (Sarma, 2008). SWIFT considers the virtual identity as the communication end point (Figure 11.2).

11.6 Future Development

The future technology developments in the IoT will focus on identification, which includes identity management using open frameworks, soft identities, or virtual identities and then transition to a "Thing/Object" DNA identifier, which is context aware but anonymous and can support multiple methods using one single unified ID. The architecture will also transition from the current focus network of network to cognitive and distributed architectures that are location aware, energy aware, and capabilities aware and are knowledge sharing. In the near future, cross domain application deployments will be in focus, which will become more global and general purpose with time and mobile applications with bio-IoT-human interaction coming into existence. In the applications front, configurable IoT devices and focus on environmental capabilities, manufacturing, and healthcare will lead to developments in the IoT wearables, application capable of high-volume data to more cluster formations like Internet of Energy, Internet of Vehicles, or Cognitive Internet, etc., and cognitive platforms based on artificial intelligence and with event-driven data acquisition. Communication technology will focus from wide-spectrum-based protocols to time sharing and unified protocols, multifunctional low-power highly configurable chipsets, and low-power short-range networks. Cognitive networks will develop, which are self-learning and self-healing running on IPV6 and supporting things to human collaboration. Swarm intelligence and adaptation mechanisms will be inbuilt into these systems. Hardware generally will evolve using nano technology and other materials including biochemical sensors useful for ingestibles. Future IoT devices will be capable of autonomous search with semantic-based discovery of sensors. Power cells will focus on biodegradable components, nano power processing units, and power generation using renewable components like movements or solar. Security will evolve to becoming self-adaptive and self-managing with privacy aware data processing. Interoperability will get supported with plug and play IoT devices and agile and operative self-adaptable methods. Standards for autonomous communication for heterogeneous networks and machines will also evolve in the future.

11.7 Conclusion

The future of the Internet of Things will be very different from what we are experiencing today, since the technology is currently still evolving and, hence, the adoption rate is still not high due to cost barriers and missing technology maturity. The issue of green design with so many billions of the IoT objects communicating and consuming energy, and limited studies on the impact on the environment on account of the need for recycling so many electronic components on one hand, and the radio signal transmission on the other, will definitely need to be addressed in the future IoT roadmaps. Again, privacy and security issues, which today have taken a backseat to technology adoption and business use case, will gain more traction. Supply chain interactions will be automated, information driven and real time with the implementation of the IoT technology. Since the IoT will necessarily need to evolve with horizontal and vertical collaboration between departments of a firm and between suppliers and customers, ecosystem models will also evolve with time. Today, organizations face immaturity of technologies, and services and vendor capabilities. The future will architect for these and manage the risk component, as new business models, approaches, and solutions start taking shape. The adoption of the IoT will most probably follow the traditional diffusion model from business leading to consumers as compared to the consumer-initiated and led model of mobility and social media. Finally, the concept of virtual factories and its application in the manufacturing domain will become widespread, which will integrate the major subsystems and can serve as a test bed for new product launches and help manage dynamic customer requirements to generate lot sizes of one, which will be basically tailor-made applications. However, there will be multiple technology barriers including energy sources to address, in which government organizations, businesses, and standards bodies will have to come together and cooperate.

References

Ali, D. S. 2017. Blockchain for IoT security and privacy: The case study of a smart home. *IEEE Percom Workshop on Security Privacy and Trust in the Internet of Thing*. Brisbane, Queensland: IEEE.

Erricson. 2013. *Erricson Energy and Carbon Report on the Impact on Networked Society*. Erricson.

FCC. 2002. *First Report and Order- Revision of Part 15*. FCC E.

Hanan, A. 2017. Internet of Things in higher education: A study on future learning. *Journal of Physics: Conference Series*, 892, doi: org/10.1088/1742-6596/892/1/012017.

Harper, R. 2010. *Warehouse Technology in the Supply Chain Management Systems*. Florida: Florida Insitute of technology.

Kosba, A. A. 2016. The blockchain model of cryptography and privacy-preserving. *Security and Privacy (SP) IEEE Symposium* (pp. 839–858). IEEE.

Lopez Research, L. 2013. *An Introduction of Internet of Things (IoT)*. San Francisco: Lopez Research LLC.

Marcel, P. A. 2017. Development of an ecosystem for the realisation of IoT services in supply chain management. *Electronic Markets*, 175–189.

McHenry, M. T. 2006. Chicago spectrum occupancy measurements and analysis. *Proceedings of the 1st International Workshop on Tech and Policy for Accessing Spectrum* (pp. 1). Chicago: ACM.

Moore, J. F. 1993. Predators and prey A new ecology of competition. *Harvard Business Review*, 71, 75–86.

Olson, M. 1965. *The Logic of Collaborative Action*. Cambridge: Harvard University Press.

Ramakrishnan, R. G. 2017. Innovation in product design: IoT objects driven new product innovation and prototyping using 3D printers. In I. Lee (Ed.), *The Internet of Things in the Modern Business Environment* (pp. 189–209). Illinois: IGI Global.

Sarma, A. E. 2008. Virtual identity framework for telecom infrastructures. *Wireless Personal Communications*, 45, 521–543.

Sivakumar, D. 2017. A case study review: Future of Internet of Things (IoT) in Malaysia. *International Journal of Information System and Engineering*, 5, 126, doi: 10.24924/ijise/2017.11/v5.iss2/126.138.

Weiser, M. 1991. The computer for the 21st century. *Scientific American*, 94–104.

12

Governance, Security, and Privacy

12.1 Introduction

The Internet of Things (IoT) architecture in applications involves the use of sensors and actuators. These provide data about environmental variables relating to movement, speed, and pressure. These further work with actuators that translate digital signals to mechanical or electrical actions in control physical systems, such as operating power circuits, applying brakes, releasing safety or pressure valves, switching on lights, or automated drug dispensers. From a security and privacy standpoint, this entire integration involves personal data about individuals and organizations on the one hand and, on the other, they control critical infrastructure remotely such as releasing pressure by activating valves or activating fire control systems. From an industry standpoint, Libelium lists more than 60 applications of the IoT with sensors and use cases. From an academic standpoint, multiple researchers have classified applications into transportation logistics, healthcare, smart environment, and futuristic opportunities (Atzori, 2010). With such use cases, there is an indispensable need and challenge of ensuring data security and data privacy. Notably, security risks are varying across interdomain levels with the latest ones like connected vehicles, health wearables and sensors, Industry 4.0 participants, smart grid and energy distribution systems, smart offices, and retail having the highest risk factors. Another challenge lies in the small form factor and limited power of devices. These devices are characterized by limited memory; they are energy efficient but have less clock cycle frequency processing capacity and need to conserve battery power by turning on and off on demand. Consequently, traditional IT security may not be suitable for these unique security challenges. The IoT devices operate on wireless networks, and are susceptible to eavesdropping. Also, with limited work having being done in IoT security, many devices are not hardened or secure and are prone to simple attacks. Data gathered by IoT devices usually reside in big data appliances on the cloud and this data is usually available to the OEM, which may give them more control over their customers leading to possible unethical repercussions. A new challenge is also due to the need for interoperability between devices supporting vendor specific proprietary protocols, so a single security solution may not work for all the IoT devices and there will be thousands of different types of devices, mostly proprietary. Finally, there is an emotional context to the IoT as humans may encounter cases where machines depose against them with proof of misuse such as car accidents or work place mistakes leading to a trust deficiency. Security threats can arise in three points where data is handled: at the point of capture, at the point of services by denial to authorized and legitimate users, and at the point of data storage achieved by unauthorized tampering or manipulation of data. The first types of attacks or "capture attacks" are meant to get access to data or get access to physical systems for further exploitation. Other class of destruct attacks are meant to immobilize the systems or damage them from functioning. Manipulative attacks aim to change data, making them meaningless or changing identity of the connected assets. The kind of hacking attacks faced by regular IT systems are also extendable to IoT systems, be it IP or identity masquerading (changing identity to provide access to some unauthorized systems such as cloned beacons or devices), man in the middle attacks (redirection of traffic with manipulation), denial of services ("DoS" aims at preventing legitimate systems users from meaningful access by overloading the system's compute power or memory or network or other resources) and distributed DoS ("DDoS" attacks are orchestrated from multiple sources on a single target). The IoT security requirements include ensuring confidentiality, availability, integrity, and authenticity. Advancements in security in IP networks such as cryptography, encryption (3DES, AES, Blowfish), transport layer protocols (TLS/SSL), or PKI-based digital certificates for private and public keys are still not implemented successfully or to the extent required in the IoT networks primarily due to the need for low computing footprint requirements since traditional IIoT networks were not designed to be open interfaces. A higher CPU consumption also requires higher power consumption, which would make the IoT endpoints not feasible. With encryption comes a higher bandwidth requirement to transmit salt code and encrypted strings, which would further compound network complexity in billions of connected IoT devices. All these point out a compelling need to relook into protocols for security, especially tailor-made, keeping in mind the IoT feasibility mandates such as low compute and memory footprint, low bandwidth, and lightweight algorithms consuming low power. Privacy concerns are more applicable for the consumer of the IoT as objects start controlling and collecting data related to our health, our homes, and environmental surrounds in a hyperconnected environment. The huge amount of data that is generated from location, vital health parameters, usage of gadgets, and electronic equipment, and something as simple as our daily life schedules from time of waking up to time to sleep, has lot of privacy challenges. Data can be collected and used covertly to monitor and label consumers based on their usage patterns, thereby removing the concept of anonymity of preferences, which is today the prospective consumer's most potent weapon against suppliers. A high degree and extent of profiling, which can be obtained by deeper analysis of the gathered data, is possible and accurate, which is particularly frightening. Even with privacy conditions and terms of agreements, it is almost impossible for end consumers to fathom the extent their data can be put to use for profiling and also whether the data can be shared not only with the original recipient but also with government bodies, remarketers, and other sellers. It becomes imperative for every organization to ensure and guarantee customer privacy and engineer systems that can implement authorization tools for storage and transit data, which can be based on any of the prevalent principles such as CIA (confidentiality, integrity, and availability), or the five pillars of information assurance (availability, authenticity, integrity, confidentiality, and nonrepudiation).

12.2 Classification

The popular context to qualify technology security is to categorize them into threats, vulnerabilities, and risks. Threats come either as manmade or natural. In the IoT context, failure of operating environment due to natural calamities can be caused by rain, fire, flood, or seismic. The IoT devices are subject to the same physical disruption and damage as other IT systems. This can cause degradation of services other than tampering of data, and can affect in CPS the interfaces that drive the physical world such as state estimation filters or sensors used for feedback and as controllers. Vulnerabilities or weaknesses are caused due to design and deployment architecture weaknesses and examples can be improper casing that can be opened and modified or software quality issues not using encryption allowing any attacker to write data to an IoT device chip and cause exploits like backdoor Trojans to be installed as root kit for remote access. Risk defined as the exposure to loss may need to be reduced specifically in an IoT context to have minimal residual risks associated with an implementation.

12.3 Types of the IoT Attacks

The following types of the IoT attacks have been observed in the recent past across different domestic and industrial applications:

- Denial of Service (DOS and DDOS)
- Physical tampering of devices
- Spoofing identity

- Protocol attacks based on the proprietary stack weakness
- Application attacks—middleware, database
- Embedded operating system attacks and crypto tampering attacks
- Access control attacks leading to privilege escalation

The potential for attacks in the IoT devices is much higher due to their distributed nature and sheer number of devices with a small footprint. It becomes mandatory to secure all end point devices since any compromised device has the potential to disrupt the entire IoT network within and outside the enterprise, causing financial loss and information loss. A use case of health-related IoT gadgets being tampered with can endanger human lives. In the case of infrastructure and utilities like the telecom, engineering, or power sector, this can lead to disruption of key communication and power networks. There is a pressing need to formulate regulatory and legal guidelines by an international independent organization ensuring the IoT architectures are resilient to denial of service focused disruptive attacks, ensure uniform authentication standards, and ensure access control in a heterogeneous environment and implement client privacy by obfuscating data. In such an ambient environment, as in a sensor-connected IoT world, the user becomes ubiquitous along with devices and their context become transparent. Having every object interconnected to the digital network creates new security and privacy problems including data confidentiality, information authenticity, and application integrity in a universal ecosystem of these objects and people. The following attributes are unique to IoT applications, making additional security parameters necessary.

- *Mobility*: The IoT sensors are mobile especially when integrated in a product and connect using a heterogeneous set of protocols, this makes them prone to physical tampering. Being randomly assigned IP identifiers based on their connectivity, it makes IP-based security mechanism difficult.
- *Wireless*: The IoT devices still implement clear-text data transmission techniques making them vulnerable to eavesdropping and interception.
- *Fixed Use*: User profiling is another risk since devices have fixed usage hence patterns can be derived based on data and location.
- *Diversity*: The IoT devices have varied computational abilities from mobile PCs to low-end RFID tags. Privacy designs must be sufficient and fulfilling for even the simplest of devices and the complex ones. Protocols must be device and vendor agnostic to be successful in implementing end-to-end security.
- *Scale*: These devices are small-form factor having IP capability and, with new developments daily, growing in numbers and reach.

The devices will count in billions and hence each of them becomes a potential security risk end point. All devices need to be hardened and patched, otherwise one loose link may compromise the entire connected network leading to financial and data losses.

12.4 Current Protocols and Challenges

Previous literature (Bonetto et al., 2012) has explained the security constraints in low compute devices explaining how elements interact. There are numerous IoT protocols like ZigBee, 6LoWPAN (Low-power WPAN using IPV6), constrained RESTful environments (CoRE), and CoAP (constrained application protocol for constrained devices supporting low power and lossy networks). Supporting security protocols, IKEv2/IPSec and TSL/SSL help implement identity management, privacy, and trust in the IoT networks. Applications of existing protocols have serious limitations in a constrained application stack (Heer et al., 2011). 6LoWPAN enables sensors to communicate directly with IP protocols over a native layer. New generation protocols such as CoAP and Message Queue Telemetry Transport (MQTT) ensure that bandwidth is optimally used. One of the key considerations for the IoT devices (Hunkeler et al., 2008), these protocols also optimize using compression to fit into Maximum Transmission units (MTU) and thereby reduce power consumption. Protocols in an IoT Bluetooth stack uses CoAP and MQTT in the application layer, IPv6 in the network layer, and Bluetooth 6LoWPAN in the adaptation layer. However, most of the IoT devices are still protected by weak passwords and technology advancements and password policies cannot be implemented in these systems, for example, a Wi-Fi access point, even today, does not have password expiration rules or complexity rules due to their limited storage and compute power. Also, a factory reset can be done by physically reset, reverting the device to its default factory settings and thereby opening up access to unintended users. Most of these devices, including Industrial Internet of Things (IIoT) devices, generate plaintext data when connected to their COM or serial ports, without authentication. Web-based admin consoles provided by the IoT vendors are themselves prone to numerous vulnerabilities, which may have been patched in web servers but are still to be patched to real-time operating systems. This problem is compounded by the focus on consumerization of the IoT devices by adding more and more business applications rather than the focus on a security centric development framework. The following technology components of an IoT implementation need to be addressed from the security viewpoint:

- Sensors, actuators
- Gateways for data transmission

- Connectivity in terms of data networks – LAN, WSN, WAN, MPLS, P2P
- Services in cloud or premises to consume the data
- Web interface to view reports and dashboards
- Mobile interface to view reports and dashboards

The IoT devices usually have a static profile that stores information about its internal state such as its operating system versions, patch set level, memory size, or computation capabilities. These devices may be vendor preset or reconfigurable by the user. The typical factory setting is vendor present; however, some application and devices support one-time update or many times update by the consumer. Security in such devices may be difficult to ensure using cryptography because of the lack of support for secure channels.

12.5 Network Level Security

The basic IPSec protocol uses encryption and authentication in a unidirectional manner, using a pre-shared key and hashing algorithms powered by the Internet Key Exchange—IKE protocol. However, IKE is too resource intensive for low compute power devices like the IoT devices. Also, data overhead is another concern while using IPSec for IoT communication because of the header information that will need to be compressed. In the CoAP architecture, DTLS protocol is proposed for the IoT security riding on top of UDP and not TCP. Similarly, 6LoWPAN incorporate headers compression and data compression with two-way authentication schemes. The IoT also requires non-IP-based devices to communicate and hence IoT gateways come into the picture. Non-IP network nodes connect to the IoT network using IoT gateways using PAN technology or directly using WAN connections. Hence, gateways need to provide adequate security for the connecting nodes. This has to be not only at the physical layer, but also providing cryptographic algorithms, and secure protocols using TCP or other adaptations optimized for the low power and compute requirements, an example of which can be seen in the Routing Protocol for Low Power and Lossy Networks (Winter et al., 2012.). The common attacks in the IoT devices are "selective forwarding attacks (SFA)," "sinkhole attacks," "hello flood attacks," "wormhole attacks," and "clone id and Sybil attacks." SFA attacks focus on disrupting services by forwarding packets from compromised nodes by selectively forwarding a specific set of messages. Sinkhole attacks, false routing paths, are advertised by compromised nodes and all traffic is routed through the node that can be than tampered with. In the hello flood attack, new nodes join in with a false "hello" message broadcast, which then becomes paired with multiple nodes. Wormhole attacks allows forwarding packets at a faster rate and can be minimized by segmenting the network. A clone ID attack is similar to an IP spoofing attack where the node identity is stolen or cloned.

12.6 Authorization and Authentication

All IoT nodes have to be authorized to join the network and for this, a process of handshake of credentials, using remote server authentication protocols such as PANA (Forsberg et al., 2008). If it is successful, higher layer security associations can be implemented using IPSec traditional Web 2.0 authentication technologies like OAuth, which rely on resource owner (grant access), resource server (throws a challenge when a protected resource is accessed), client or consumer of services, and an authorization server that issues tickets or tokens. However, this cannot be implemented in the IoT due to the resource constraints. Hence, the authentication scheme is delinked from the IoT devices and federated to an external agent IoT-OAS, which handles the issue of tokens (Cirani et al., 2015). Concepts such as SSO (Single Sign On), which allows legitimate users to authenticate only once against services before being provided access to multiple resources that share a federated security-trust-based architecture, based on Shibboleth OpenID and OAuth2, are still not fully compatible with the IoT devices.

New security trends for the IoT devices focuses on lightweight cryptography that can be obtained in a constrained compute environment and can take the shape of symmetric or asymmetric ciphers. While the former is used for message integrity checks, the latter is used for key management and is more resource intensive. Commonly used symmetric key algorithms include Tiny Encryption Algorithm (TEA; Wheeler & Needham, 1995), which focuses on numerous iterations using 32/64 bit signed integers for XOR or shift operations and Scalable Encryption Algorithm (SEA) used in small embedded applications using a plaintext key of a limited key size (Standaert et al., 2006). Heterogeneity in the IoT stand devices implies that some devices may be operating on a partial protocol stock and may lead to security in obscurity due to developer-implemented security programs. There is a trade-off between anonymity and accountability and pseudo-anonymous systems will be required linking participants to a random generated identifier than an actual identity, which is also a challenge in the IoT. It is also necessary to have access controls at the end devices, more so in a fog computing scenario, using a local access controller (Cerf, 2015) to identify whether a one-time authenticated user has access to resources on a continuous basis (Abomhara, 2014). Access control can be implemented either in role based, discretionary, or attribute based, all of which are challenging in a low compute power or decentralized environment, hence, capability-based access controls are being preferred for the IoT environment.

Security must not make technology adoption unfeasible and should be low cost, reliable, and manageable. In an IoT environment where the targets may be geographically spread out, detection of vulnerabilities through logs and putting appropriate patches must be an ongoing process. This has led to a rise of remote access and cloud-based patching methods, which also have the risk of altering system functionality, which may be or may not be as per user expectations and, hence, may be subject to user choice on whether the patch is applied, or not if it requires some sort of user confirmation or intervention. There have been some cases of botnet attacks using unpatched printers, IP cameras, etc., such as in the Dyn 2016 attack. Such attacks can have serious legal liabilities and generate customer-negative sentiments eroding brand reliability. Various end-point diagnostic devices such as OBD and DSRC can be exploited easily in the case of autonomous or driverless vehicles (Checkoway, 2011), where vehicles could be controlled when in motion using exploitation of unpatched vulnerabilities. Similarly, in the case of healthcare, the MEDJACK attack impacted healthcare equipment like blood gas analyzers and x-ray machines leading to leaked medical data or worse, the disruption of services by tampered readings, which could cost human lives. Disruptions to Ukraine's energy networks, the Iranian nuclear industry, and incidents of variations of ship coordinates have surfaced in the recent past. Also, smart offices and their HVAC systems have been compromised to stall competitors or prevent normal operations.

12.7 Privacy in the IoT

Privacy is a major issue with the IoT as enormous data becomes available of citizens, governments, and consumers. Trust is the main element in the adoption of any new technology and the same goes with the IoT. There is a lot of difference across age and education band users when it comes to removing digital footprints, even when browsing the internet. With microlevel-embedded sensors transmitting data every second with details of an entity and its surroundings, it becomes very easy to identify any entity based on data extrapolation specifically in healthcare, consumer preferences, legal enforcement, and other sensitive domains. Fear of privacy invasion can act as a deterrent to adoption of IoT technologies in business and consumer areas. Data anonymization and obfuscation should be used to delink the identities of entities from the data using one-way encryption and enforcing digital shadows or virtual identities. A proven area of homomorphic encryption such as an RSA algorithm can find utility in IoT networks. This involves operations on ciphertext giving cipher results that are same as transactions done on plaintext. Hence, source data can be encrypted to ensure privacy even from privileged system administrators and yet perform transactions on the ciphertext. This will result in lower power consumption being used to decrypt at each stage of storage and transmission, which is in sync with the IoT objective but the downside is that to have a comparable level of security as traditional networks, a lot of computing power may be needed. There have been some developments in each stage of homographic encryption, that is, encryption at the transmitting node before starting to transmit data, decryption at the receiving node, aggregation of data, and then final encryption before further storage. This can be something like the project Enigma (Zyskind et al., 2015), which is a peer-to-peer network allowing a joint multiparty store and compute facilities for private data. The compute is based on optimized yet secure multiparty secret sharing scheme, the storage is based on hashing using secret shared data, and it also implements blockchain for tamper proofing audit trails, thereby ensuring autonomous control with cryptographic security. Another technique in the IoT can be zero-knowledge protocols, which is a substitute of public key protocols and differs in the way key exchanges can happen without sharing any secret information and works at less than a hundredth of the compute power. Privacy and personalized services have to be weighed and balanced as per consumer preferences, and, hence, consent of users is necessary before capturing or storing their data and that too for specific purposes, which should be shared transparently. In the case of the IoT, interaction will be with devices not having a physical display or console and, hence, it becomes a bigger challenge in ensuring the user gives consent to data without seeing what exactly is getting captured and whether subsequent patches are also not capturing data more than what has been consented for in the first place. Smart toys have the ability to monitor and recognize the voice of children and with their end points powered by Bluetooth or Wi-Fi are susceptible to intruders to monitor. Apart from the dark side, smart equipment like refrigerators can provide insight on eating habits to insurance companies or even cars can become source of witness against the car owners in case of an accident. Privacy concerns also affect the IIoT, due to numerous attack vectors and exposed surfaces, and this can lead to loss of intellectual property and open up confidential data to competitors.

12.8 Governance Issues

The scale and complexity of the IoT deployments, coupled with new technologies like the cloud and big data, gives rise to lot of governance issues. The IoT governance refers to shared principles adoption in a multiparty IoT ecosystem, which may involve government, private players, service providers, and consumers. The main objectives are to ensure addressability and unique identification, security and stability, ensure free and fair competition, and protect data from unauthorized use. Identification relates to the network address and device identifier, and providing interoperability to ensure an Internet, and not Intranet, of Things, which are separated by integration challenges. Privacy, while at the same time ensuring data owners do not restrict the flow of data, is another aspect that can be ensured in either a regulatory framework or technologically. The rights of people to privacy is the ability to have autonomous decisions and control their networked environment. Focus on decentralized solutions offering more autonomy and security has become a requirement with concepts such as smart grid and smart cities. The concept of the Internet of Everything (IoE) brings together data, process, people, and objects and, hence, governance has to cover modalities for catering to all of these individually and as an integrated unit since they bring together services, intelligence, environment, and context (Bojanova, 2014). The IoT has evolved from different technologies over multiple visions of different communities with varying interests using proprietary standards, this has led to lack of standardization, creating issues of data security, data interchange, and privacy since corresponding controls and checks have not evolved making security risk analysis and risk mitigation more complicated.

Governance objectives in the IoT are not distinctly mapped to the operations process, giving rise to a situation where business stakeholders (in contrast with traditional technologies) do not form part of governance objectives, which are much broader in domain. In the case of IIoT, there are multiple stakeholders—primarily, the manufacturer, the distributor, end customer, and the raw material supplier—each having their own set of business factors, for example, the manufacturer is concerned with production and quality, the raw material supplier is focused on their product quality and compliance, the distributor is also focused on customer satisfaction and servicing, while the end customer is primarily concerned with the value they derive. The IIoT setup either happens at the manufacturing asset level (industrial) or the end product level (domestic) are is largely distributed. Governance aspects can be identified as one of environment centric, data centric, or infrastructure specific. While the first deals with overlap of controls between multiple players and how it can be segregated, such as in the case of energy management systems, the second is more to do with privacy, security, and trust issues, while the last one deals with configuration and technical deployments of the supporting infrastructure. Similarly, governance of operational processes can be grouped as either configuration centered, topology centered, or stream oriented. The first deals with runtime configuration and capability management, for example, provisioning services on demand basis, the second deals with pushing changes in the technology stack in the IoT devices, while the last deals with manipulation and reporting of data being gathered in streams from end-point IoT devices. Traditional approaches that work on assumptions of getting remote access, monitoring through logs, or pushing configuration are hardly feasible when we talk of billions of connected devices in an IoT world and hence ad hoc practices are being adopted in operational execution of the IoT projects. Also, governance factors such as legal practices in different countries and their security or privacy policies may require operational decisions such as data to be stored in different locations. Due to the complexity of implementations, it is also necessary to protect stakeholders from the underlying challenges that they may not be aware of or in a position to decide what is best suited for them. Factors to be considered in the IoT governance include unified enforcement capabilities, dynamic and on-demand governance capabilities in a large-scale deployment, precise and yet flexible controls, declarative policies that can take effect at runtime, and higher degree of autonomy to devices. Governance capabilities could be explored as microservices, which can be added or removed at runtime by remote provisioning initiated by the devices themselves rather than requiring user or technical inputs. There have been many developments in the IoT governance including security and privacy with legitimacy (Weber, 2013), data quality issues (Weber, 2009), and operations management including infrastructure virtualization and provisioning (Soldatos et al., 2012).

Another area of governance involves the inclusion of ethics elements in the IoT governance as supported by public consultation such as identity, fairness, and autonomy of people. An example of using the IoT data on demographics to give higher priority to data of a few individuals for medical treatment would not be considered ethical. It may be difficult to enforce this only on the basis of consent unless suitable regulations at sovereign level applicable across countries are drafted with strict penalties. Interoperability is also necessary to promote competition and prevent vendor lock-in in services. Unique identification through globally enforced schemes, as in the case of IPv4 and IPv6, would also be needed for non-IP-based IoT devices. There also seems to be a disconnect between countries on whether the existing internet governance practices are sufficient for the IoT or new safeguards and agencies are separately needed. To substantiate this, we need to identify some of the peculiarities between the IoT and the internet, first being the IoT is driven by object name services (ONS) based on EPC code identifiers while the internet was driven by domain name services (DNS). ONS uses the EPC global standards while DNS follows standards laid down by RFC (request for complement series). The naming schemes also differ: DNS can have up to 63 octets alphanumeric character names while the ONS uses Tag Data Standards, and the DNS model is based on public infrastructure models while the ONS is more private RFID-specific infrastructure. Hence, the internet focuses on domain and their registrations while the IoT will focus on identifiers and their attributes.

12.9 Conclusion

With IoT, security and privacy issues are more prominent than the World Wide Web where business models like social media are built on the concept of information being stored and then the data being used for cross selling. With concepts like wearables, the IoT devices store much more sensitive information such as healthcare, or user preferences. Similarly, due to the physical small-form factor of the IoT devices and their ubiquitous nature providing anytime anywhere compute, security challenges arise from any vulnerable end points. In the case of IIoT, devices such as plants or assemblies traditionally designed to focus more on operations technology (OT) than information technology (IT) have far limited compute power, which makes traditional security algorithms and methods useless. Redesigning many of these systems, which do not even operate on an IP stack, requires introducing a layered structure of devices, thereby making security issues even bigger. This will involve relooking into authentication and authorization schemes and trying to establish feasibility of many research propositions. This holds promise but needs to scale up for practical usage. Security solutions must be capable of being used by different types of users across demographics and age and support heterogeneous platforms and be vendor agnostic. There should be certifications for the IoT services. The IoT governance remains one of the key challenges, and is critical for success in adopting projects. Given the multitude of devices, services, vendors, and partners involved in an IoT project, an inclusive governance framework is a requirement. However, in the future, a lot of initiatives in security, governance, and privacy, which are still in laboratory stage, are expected to become commercially viable and address pertinent issues.

References

Abomhara, M. A. 2014. Security and privacy in the Internet of Things: Current status and open issues. *International Conference on Privacy and Security in Mobile Systems (PRISMS)* (pp. 1–8). Aalborg, Denmark.

Atzori, L. A. 2010. The Internet of Things: A survey. *Computer Networks*, 54(15), 2787–2805.

Bojanova, I. G. 2014. Imagineering an internet of anything. *Computer*, 47(6), 72–77. doi: 10.1109/MC.2014.150.

Bonetto, R., Bui, N., Lakkundi, V., Olivereau, A., Serbanati, A., Rossi, M. 2012. Secure communication for smart IoT objects: Protocol stacks, use cases and practical examples. *IEEE International Symposium on a World of Wireless, Mobile and Multimedia Networks* (pp. 1–7). San Fransisco: IEEE.

Cerf, V. G. 2015. Access control and the Internet of Things. *IEEE Internet Computing*, 19(5), 96. doi: 10.1109/MIC.2015.108.

Checkoway, S. D. 2011. Comprehensive experimental analyses of automotive attack surfaces. *2011 USENIX Security Symposium* (pp. 77–92). San Francisco, CA.

Cirani, S., Picone, M., Gonizzi, P., Veltri, L., Ferrari, G. 2015. IoT-OAS: An OAuth-based authorization service architecture for secure services in IoT scenarios. *IEEE Sensors Journal*, 15(2), 1224–1234.

Forsberg, D., Ohba, Y., Tschofenig, H., Yegin, A. 2008. Protocol for carrying authentication for network access (PANA). *RFC 5191, RFC Editor.*

Heer, T., Garcia-Morchon, O., Hummen, R., Keoh, S. L., Kumar, S. S., Wehrle, K. 2011. Security challenges in the IP-based Internet of Things. *Wireless Personal Communications*, 61(3), 527–542.

Hunkeler, U., Truong, H. L., Stanford-Clark, A. 2008. Mqtt-s—A publish/subscribe protocol for wireless sensor networks. *Communication Systems Software and Middleware and Workshops, 2008. IEEE COMSWARE* (pp. 791–798).

Soldatos, J., Serrano, M., Hauswirth, M. 2012. Convergence of utility computing with the internet-of-things. *2012 Sixth International Conference on Innovative Mobile and Internet Services in Ubiquitous Computing* (pp. 874–879).

Standaert, F.-X., Piret, G., Gershenfeld, N., Quisquater, J.-J. 2006. SEA: A scalable encryption algorithm for small embedded applications. *Proceedings of 7th IFIP WG 8.8/11.2 international conference, CARDIS* (pp. 222–236). Tarragona, Spain.

Weber, K. 2009. One size does not fit all—A contingency approach to data governance. *Journal of Data and Information Quality*, 1, 1–4.

Weber, R. H. 2013. Internet of things–governance quo vadis? *Computer Law & Security Review*, 29(4), 341–347.

Wheeler, D. J., Needham, R. M. 1995. TEA, A tiny encryption algorithm. *Proceedings of Fast Software Encryption, 2nd International Workshop* (pp. 363–366). Leuven, Belgium.

Winter, T., Thubert, P., Brandt, A., Hui, J., Kelsey, R., Levis, P., Pister, K., Struik, R., Vasseur, J. P., Alexander, R. 2012. RPL: IPv6 routing protocol for low-power and lossy networks. *RFC 6550, RFC Editor.*

Zyskind, G., Nathan, O., Pentland, A. 2015. Enigma: Decentralized computation platform with guaranteed privacy. *CoRR*, abs/1506.03471.

Index

For Product Safety Concerns and Information please contact our EU
representative GPSR@taylorandfrancis.com Taylor & Francis Verlag GmbH,
Kaufingerstraße 24, 80331 München, Germany

Printed and bound by CPI Group (UK) Ltd, Croydon, CR0 4YY
08/05/2025
01864366-0019